Wetland Birds

Wetland birds provide us with some of nature's most wonderful sights, from vast flocks wheeling overhead to newly hatched chicks drying in the sun. Apart from their beauty and recreational and economic importance, these birds are excellent indicators of water quality and measures of biodiversity. But how do they use wetland habitats, and how can we best conserve and maintain them for the future? Here, Milton Weller describes the ecology of wetland birds by identifying patterns of habitat use and typical bird communities that result from the use of resources such as food, cover, and breeding sites. He integrates basic and practical information on bird–habitat relationships for researchers, landowners, managers, and birders alike. As wetlands continue to decline, this book will help us to understand the potential and the limitations of wetlands as bird habitats.

MILTON W. WELLER is Professor Emeritus at the Texas A & M University. His research interests have emphasized wetland bird–habitat dynamics, waterbird behaviour, bird communities and wetland restoration and management. He is author of *The Island Waterfowl* and *Freshwater Marshes*. His many honours include the Gulf Conservation Award, the Lifetime Achievement Award from the Society of Wetland Scientists, and the prestigious Aldo Leopold Award Medal from The Wildlife Society.

Wetland Birds

Habitat Resources and Conservation Implications

MILTON W. WELLER

CAMBRIDGE
UNIVERSITY PRESS

PUBLISHED BY THE PRESS SYNDICATE OF THE UNIVERSITY OF CAMBRIDGE
The Pitt Building, Trumpington Street, Cambridge CB2 1RP, United Kingdom

CAMBRIDGE UNIVERSITY PRESS
The Edinburgh Building, Cambridge CB2 2RU, UK   http://www.cup.cam.ac.uk
40 West 20th Street, New York, NY 10011–4211, USA   http://www.cup.org
10 Stamford Road, Oakleigh, Melbourne 3166, Australia

First published 1999

Typeset in Adobe Minion 10.25/12.75pt. *System* QuarkXPress ®   [SE]

A catalogue record for this book is available from the British Library

Library of Congress Cataloguing in Publication data

Weller, Milton, W.
Wetland birds : habitat resources and conservation implications /
Milton W. Weller
    p.   cm.
Includes index.
ISBN 0 521 63326 5 (hb). – ISBN 0 521 63362 1 (pb)
1. Water birds – Ecology.   2. Wetland animals – Ecology.   3. Birds,
Protection of.   I. Title.
QL698.95.W45   1999
598.1768–dc21   98-21973   CIP

ISBN 0 521 63326 5 hardback
ISBN 0 521 63362 1 paperback

Transferred to digital printing 2001

To many former students who are doing so much for the conservation of wetlands and wildlife, especially to BOB BERGMAN and JOE MOORE who gave their all

Contents

Plates

Preface

Wetland birds have long attracted the attention of the public and scientists because of their beauty, abundance, visibility, and social behavior, as well as for their recreational and economic importance. Recently, they have become of interest as indicators of wetland quality, and as parameters of restoration success and regional biodiversity. Recognition of the importance of wetland habitat resources as the dominant influence on the presence and repeated use of wetlands by birds has come slowly, but sufficient information is now available that it seems timely to summarize what we know, and to speculate on what we think we know, as a means of seeking direction for future work. This book is partly a product of my own interest in patterns of wetland habitat use by birds in relation to wetland dynamics. In addition, nonspecialists have been seeking help in using birds as indicators of wetland function or as measures of success in wetland management, restoration, and creation. I have tried to make the book understandable to students of various ages and stages who are interested in wetland birds and their conservation, but some may find more detail than they like. Others with greater knowledge of certain topics will find it advantageous to skip from the known to the less well-known subject matter. The literature on wetland birds now is enormous, but to avoid constant interruption of the text, I have cited literature in some chapters less than some may prefer. However, some chapters list further reading material not cited in the text; in both sections I have tried to include classics in the subject area and more recent papers or books that provide summaries of pertinent literature. Words or phrases that are important to the understanding of the topic and that are in common use in the field are in **bold-face** type and usually are defined directly or by example.

The focus of the book is on how wetland habitat resources influence its use by birds in general rather than on any specific taxonomic group of birds, or their breeding biology – a separate but massive subject in itself. I have used the term **wetland birds** in the title to provide a habitat focus that considers birds adapted to a wide range of relatively shallow waters through to those in wet-soil habitats. The term **waterbirds** is often used for these various groups, commonly identified by habitat- or resource-related descriptors: divers (especially loons), waterfowl (wildfowl in Europe, but also used broadly like waterbirds), waders (herons in Europe), shorebirds (waders in Europe), littoral or water-edge birds, aquatic birds, coastal birds, and estuarine birds. Some of these groups of birds avoid saline wetlands but may use coastal or inland wetlands as long as they are freshwater; others use mainly saline wetlands. Some species may use either fresh or saline wetlands but use is influenced by factors other than salinity: for example, life cycle function such as breeding or nonbreeding periods. The book does not treat species of the deep-ocean (i.e. **oceanic, pelagic,** or **seabirds**) except as they come to coastal wetlands to nest, feed, or roost in shallow and food-rich waters. The last usually nest in groups and have attracted the attention of ecologists and ornithologists, who have focused on the social behavior of **colonial waterbirds,** a term that includes many freshwater species as well. Overlap in categorizing such birds by habitat characteristics reflects both the dynamic nature of wetlands and the evolution of great adaptability to exploit wetland resources wherever they are.

To aid the interested but less experienced, I have included a chapter describing the various taxonomic groups that use wetlands, with emphasis on habitat adaptations, foods, and habitat-influenced breeding biology. The reader will find a wide range of birds, from those groups where most species are **obligate** users of wetland habitats through taxonomic orders or families where only a few species are so restricted to those that are regular users but do so **facultatively.** Obviously, almost any bird that happens to be at a wetland edge may occasionally exploit wetland resources such as water or food opportunistically. This range of variation may disturb some readers who search for simple patterns, but it reflects the dynamics of many wetland habitats and the amazing flexibility of birds. Although photographs throughout the book provide examples of activities and adaptations of various wetland bird species, the book does not deal with identification. Several bird guides covering specific taxonomic groups or geographic areas have been included in the bibliographies. To save space and simplify reading, scientific names are given in Appendix 1 for birds, Appendix 2 for other animals, and Appendix 3 for plants.

Most books written about the taxa of birds that regularly use wetlands have been taxonomically structured even when emphasizing life-history information. Moreover, habitat descriptions have been limited or lacking in many such books partly because of the absence of information on species and adaptations to resources such as food, but also because of the difficulty of finding widely recognized descriptive terms. This is rapidly changing because of current interest in the conservation of natural habitats as the essential approach to preserving threatened species and maintaining regional biodiversity. This much-needed habitat focus has increased the need for information on the environmental resources likely to be found within the habitat. Several recent books on wetland birds reflect this growing habitat interest, and hopefully this discussion as well as the publications cited will encourage further descriptive as well as integrative work.

After an introduction to habitat concepts essential to provide an equal footing for all readers, subsequent chapters of the book will outline wetland diversity and classification, review the major groups of bird that use wetlands, consider how wetland features influence bird biology and adaptation, elucidate how birds can influence wetlands, examine methods of describing potential wetland microhabitats, and identify how we might relate changes in wetland bird communities to the dynamics of habitat resources over time and space. Opportunistically throughout the text, I shall try to relate patterns to potential application in the conservation and management of wetland birds, and additional issues will be summarized in several chapters near the end of the book.

This book reflects my biases by virtue of experience with certain taxonomic groups, geographic areas, and literature, but I hope the examples reflect a view of species as part of a community and consider the features and dynamics of habitats as driving influences on the evolution of bird groupings as well as on species' attributes. For groups I know poorly, I have tried to incorporate examples documented in the literature. I suspect this will satisfy few specialists in those areas, but I hope there is enough information to allow us to focus on general patterns. Many patterns will serve only as hypotheses for future testing, which should help determine not only what occurs but how and why such patterns exist. Therefore, it is especially important for readers to evaluate general statements of apparent patterns analytically.

Asking general questions important in wetland ecology may be relatively easy – but answering them is not! What is it about wetlands that produces such concentrations of birds, both in breeding and nonbreeding periods? Unlike our expectation of terrestrial birds, why are they often present in one year and not another? Why are many of the groups so widely distributed and among the most mobile animals in existence, whereas other species have modest ranges and are so scarce that the list of endangered wetland species is long? How has this habitat influenced breeding biology and life-history strategies? Why are some birds not represented among wetland users, and what does this tell us about the wetland ecosystem? How can we apply our knowledge, even when minimal, in a conservation strategy of protection, management, or restoration?

Persons wanting to preserve wetland bird communities, enhance bird populations, preserve species' diversity through management, or to use birds as an index for assessing habitat quality usually want to know which single habitat feature is most important to birds, how data can be gathered quickly, and how this knowledge can be applied in a simple and practical way. Unfortunately, ecological problem-solving is neither simple nor conclusive. The answer requires not only a significant amount of life-history information about each bird species or group in question, but consideration of the quantity, quality, and dispersion of biological resources such as food, vegetation, and other animals; additionally, a knowledge of physical features such as water, ice, and geomorphology are essential. Issues of **spatial scale** (size, shape, and dispersion of components) and **temporal scale** (daily, seasonal, annual, irregular but still time-related influences) complicate the picture further. Therefore, most problems are, as in all natural ecosystems, multivariate and difficult to test and to use to draw indisputable conclusions. Much of the book will emphasize understanding such difficult questions, but I suspect that I will answer too few of them. That is, however, part of the strategy of the book: most generalizations induce questions on the basis of personal experience – to the ultimate good of science. To that end, I do not hesitate to generalize with the hope that such statements result in challenge, additional insights, observation, and, ultimately, new information.

Acknowledgments

I am indebted to so many people and organizations that listing without missing some is impossible. My wife Doris has supported my endeavors without hesitation but with meaningful questions and astute observations for many years. Our son Mitchel observed and learned from a forced outdoor experience that has been reflected in his activities in later life. Many undergraduate and graduate students, both those working with me and with colleagues, have asked good questions and contributed much to my experiences and background, as I hope I have to theirs.

My work at four unique universities, the University of Missouri, Iowa State University, University of Minnesota, and Texas A&M University at College Station, was supported in concept and salary if not always in research funds. But many other organizations assisted in financing, program support, and field opportunities, guided by people who shared interests and data needs: US National Science Foundation, US Fish and Wildlife Service, US Geological Survey, US Environmental Protection Agency, Delta Waterfowl Research Station, Wildlife Management Institute, National Audubon Society, Welder Wildlife Foundation, South Florida Water Management District, Texas Utilities Company, US National Research Council, British Antarctic Survey, New Zealand Wildlife Service, Buenos Aires Natural History Museum, The Wildfowl and Wetlands Trust, and state conservation organizations in Missouri, Iowa, Minnesota, Texas, and Florida.

Among the many persons who have been especially important in my background studies, insights, and in giving direct help related to this book are (alphabetically): Frank Bellrose, Jean Delacour, Dirk V. Derksen, Jim Dinsmore, William H. Elder, Paul L. Errington, Leigh H. Fredrickson, Harry Frith, Al Hochbaum, Art Hawkins, A. W. Johnson, Janet Kear, Gary Krapu, Morna MacIver, Frank McKinney, Peter Miles, Mike Milonski, Claes Olrog, John P. Rogers, Maurice Rumboll, David L. Trauger, Paul L. Vohs, Peter Ward, and Dick White. Tracey Sanderson, Jane Ward, Sue Tuck, and others at Cambridge University Press were extremely helpful and efficient in their support of this project. I am indebted to all ... and many more.

1

Introduction

1.1 The habitat perspective

Wetlands are unique **biotic communities** involving diverse plants and animals that are adapted to shallow and often dynamic water regimes. Here we focus on these systems as **habitats** for birds, viewing them as providers of the resources that birds need to survive and reproduce, but also as major forces in the evolution of their life-history strategies. As water is the key to wetland existence, the commonality of the birds considered here as wetland birds is their response to shallow waters, whether salty or fresh, and their use of water-based resources. Wetland birds are extremely diverse, reflecting early anatomical and physiological adaptations to this unique but rich habitat. Several birds are among the few ancient ancestors that survived whatever cataclysmic events caused the die-off of the dinosaurs (Gibbons 1997), strongly suggesting that a wet habitat was a good place to be early in the history of bird evolution, and that it still is. This attraction and response by birds to such shallow water is conspicuous and much studied; however, in addition to its inherent value to birds, water also reflects the status of other resources in a habitat at a given time. Many ecologists feel that birds are one of the more visible indicators of the total productivity of such biotic systems.

The advantage of taking a habitat perspective of a large group of diverse birds is that it helps to explain some of the ecological and evolutionary mechanisms that have dictated bird form and function. The habitat view has grown in importance but it is not new. Students of wetland birds early in the 20th century recognized the link between habitat and bird life history, but the focus often was on single species. For example, Arthur Allen's 1914 monograph *The red-winged blackbird* was subtitled, *the ecology of a cat-tail marsh*. As part of his study, he classified and mapped plant communities and stressed the interrelationships of habitat resources like food with marsh use. Many of the monographs describing species studies during the 1920s to 1960s included aspects of habitat such as food and vegetation structure, which led to better understanding of the role of water depth and dynamics in the attractiveness and maintenance of such resources (e.g., Allen 1942, Bennett 1938). Life-history studies identified many important biological attributes by species (Bent

1919–68). These ecological observations and management-oriented survey data led to the identification and listing of use of resources such as food (McAtee 1939, Martin, Zim and Nelson 1951). Lack (1933) pointed out the importance of vegetation structure used by birds as a way of identifying favored habitats. One of the first multispecies or community-oriented studies of birds in wetlands linked song perches and nest sites to wetland vegetation by mapping locations of a small assemblage of marsh birds (Beecher 1942). Beecher also analyzed bird relationships to cover–water or cover–cover patterns and noted bird response to annual changes in water regimes. Bird habitat also was an important part of the developing concept of the biome and other major biogeographic classifications (Aldrich 1945).

Many studies of bird–habitat relationships have been conducted since, and for many purposes and in many ways. There are several good reviews, conceptual papers, and analytical treatises that deal with vertebrates in general (e.g., Klopfer 1969, Morrison, Marcot and Mannan 1992, Southwood 1977), habitat selection by birds (Cody 1981, 1985, Hilden 1965, James 1971), and bird community structure with strong emphasis on habitat relationships (James 1971, Wiens 1989). Those that have focused solely on wetland birds have not only identified what bird species are present, but also how they use wetlands and how habitat features and resources influence the level of use and bird success under those conditions (Burger 1985, Weller and Fredrickson 1974, Weller and Spatcher 1965). At the same time, the role of habitat on bird pair bonds, spacing, and other breeding behavior was identified (Orians and Willson 1964, Verner and Willson 1966, Willson 1966) and linked directly to resources such as food (Orians 1980).

I suspect that few observers seeing birds in a natural habitat really ignore their adaptations for that environment, although they may not be immediately conscious of it. For example, it is not surprising that zoos are experiencing revitalization through the use of natural habitat displays of animals – often including wetlands – because visitors can see why they have the adaptations they do and, therefore, understand why certain birds are in certain places. But to fully appreciate the long-term ties of birds to wetlands, and how wetland resources and other features relate to the biological, physiological, and physical needs of birds, we must consider many viewpoints and facts that will challenge the reader to integrate and analyze these observations.

1.2 Some biological issues

Most residents of the northern hemisphere have a vision of birds, bird communities, and bird habitat strongly influenced by terrestrial species. Such "typical" birds are recognized as migratory, seasonal, and regular in routes and timing of movement. They are often loyal to both breeding and wintering habitats – which may even have similar vegetation and other features (e.g.,

Table 1.1 *Some avian adaptations to wetland habitats at two arbitrary levels*

Long-term adaptations	Shorter-term or regional variation
Anatomy and morphology	Specific foods
Rear leg placement for swimming/diving	Feeding tactics
Bone and lung modifications for diving	Feeding flights
Eye modifications for nocturnal and underwater vision	Local flight paths
Flight adaptations for aerial divers	Vegetation species and life form
Webbed and lobed feet	Nest sites and vegetation types
Long legs for wading	Water depth preference
Bill specializations for grabbing, straining, digging, etc.	Roosting patterns
Water-resistant plumage	
Behavioral adaptations of preening and drying	
Specialization for general types of food	
Physiology: respiratory, digestive tract	
Breeding strategies	
Social behavior: spacing, aggression,	
Water depth adaptations	
Wetland types, water dynamics and salinity	
Long-range mobility/resource exploitation	

Lack 1968). Brightly colored males often precede dull-colored females in migration and establish a territory to which females are attracted. Some may even re-pair with the same mate and nest again at the same site. Among terrestrial birds, the annual molt typically follows breeding, and birds retain powers of flight year-round. Such patterns are, in fact, highly variable among terrestrial birds, and especially between hemispheres, but wetland birds often differ significantly for reasons that seem habitat-related. Another influence we face is that we tend to know one species or group well but rarely compare it with other wetland birds in the same habitat; for example, although a duck or a shorebird behaves differently to a terrestrial thrush, there are also differences in habitat adaptations among the wetland birds themselves. It is the commonalities of such groups that we strive for here, as therein lies the opportunity to understand what habitat features drive their similarities and differences and how we might conserve several species or groups by treating them as part of the same system.

Avian adaptations to utilize wetland and other aquatic systems are diverse and include anatomical, morphological, physiological and behavioral changes (Table 1.1). Anatomically, they include designs for diving and swimming, such as body compression to increase specific gravity (loons and grebes), compressed body structure to allow them to pass between dense vegetation (rails and bitterns), or adaptations for plunge-diving from great heights. Visual systems must be geared to acuity and protection both in water and air, and to eye accommodation of as much as 50-fold, which allows a quick change of focus from near to far in seeking food, and to binocular vision, which occurs in herons and kingfishers (Campbell and Lack 1985). Birds that feed at night, like skimmers, have eyes that are better adapted for nocturnal vision compared with diurnal species (Rojas, McNeil and Cabana 1997). Respiratory physiology differs dramatically in those species that engage in long-term and deep diving. Although amateurs by comparison with large penguins, long-tailed ducks and loons have been recorded in fisherman's nets at 180 and 240 feet (54 and 72 m), respectively, during winter when they may leave wetland fringes for deep water (Schorger 1947). Morphologically, adaptations include bills that strain, peck, spear, store, and grab, and feet that allow swimming, diving, walking on mudflats, wading, or grabbing and holding fish. Not only do body parts differ in general form , but size of bills, legs, and flight patterns differ across a gradient of wetland edges, as noted in both North American and European peatlands (Niemi 1985). Additionally, special feathers and plumage designs ensure waterproofing under the most severe conditions.

Food types also are diverse and demand highly specialized digestive systems: some wetland birds specialize on plant diets, some on animal, some use both, and most switch foods seasonally as induced by needs for breeding or energy for migration. Water also provides protection from predators, and many waterbirds capitalize on the protection combined with the rich late-summer resources to molt all their wing feathers simultaneously. Most terrestrial birds or birds that feed on-the-wing must do this annual molt a few feathers at a time, retaining flight capability while reducing energy demands.

As a result of these adaptations, birds are better equipped as a group to exploit wetland resources – wherever they occur – than any other animals, except some mobile insects that use similar approaches but on a smaller scale. Because birds are larger, more colorful, fewer in number of species and, therefore, better known, they are more often used as indicators of conditions within a wetland ecosystem.

1.3 Some ecological viewpoints

Essential for an understanding of this book are definitions of some ecological terms that may be used in different ways by various groups or individuals. Wetlands are among the richest of **ecosystems** – a term used to define the concept of a biological system that with its interacting physical and chemical components forms a functional and self-sustaining entity. I do not use this term as a spatial or geographic unit (now common in the popular press), but there are terms like **ecoregions** (Bailey 1995) and other terms that can be used. Rather, the ecosystem is viewed as a conceptual entity or system limited more by interactions and interrelationships than by size or components (Marin

1997). Wetland richness is expressed in annual **productivity,** measured as gross primary production based on the amount of energy fixed per year (Begon, Harper and Townsend 1996, Odum 1971), and birds could not be so prominent there without the resultant resources. But, wetlands also are highly variable in productivity because of dynamic water regimes, and a single wetland probably will not have similar attraction to birds at all times within a year or among years. Such variability in resources induces mobility as birds "island-hop" to track foods in widely dispersed wetlands. Birds seemingly appraise wetland resource conditions quickly, checking known areas first, but moving elsewhere as necessary in search of more and better food, water, and other resources. Other individuals, often but not always young birds, may pioneer in areas where they may be successful in establishing viable populations – at least for the favorable years. Mobility is the key to both their individual and population survival and their reproductive success in a dynamic environment. Understanding such patterns of habitat use is essential for the development of conservation strategies, providing the basis for policy and management decisions.

The term **habitat** is not consistently used by all biologists (Hall, Krausman and Morrison 1997); I follow the traditional idea of place or space in which an organism lives (Odum 1971), assuming the presence of sufficient resources for the bird's maintenance during a portion of its annual cycle: food, water (some for swimming as well as drinking), cover (for protection from predators and weather), rest (including protection from natural and human disturbance), and space (pair-space for breeding and social- space during nonbreeding periods). The term **biotope** is used by some workers to separate the habitat of a community or collection of species from that of a single species (Whittaker, Levin and Root 1973), but this separation not been widely followed. The behavior of a bird in use of such habitats may involve aggressive defense of an area for breeding or feeding (**territory**) and occupancy of a larger, general-use area, or **home range**, where total resource needs are met but overt aggression is less obvious. There is immense variation between groups and species of birds in such requirements, and some seems to be habitat related. To clarify that habitat is not solely vegetation, the term **habitat structure** is used because this includes the physical habitat aspect of vegetation (**life form** or physiognomy of plants or plant groups plus **layers** of foliage) and also nonbiological structures. For example, a waterbird might nest on a cliff (Canada Goose), in a tree hole (e.g., Wood Duck, several mergansers, Prothonotary Warbler), an old mammal burrow (goldeneye), a dug bank burrow (kingfisher or Bank Swallow), or on a snag (Osprey), but it is the physical form rather than its biological character that is sought by the bird (and these are the species that respond to artificial nesting structures). Vegetation layering is especially prominent in forests (including forested wetlands) where various life forms of woody vegetation show dramatically different heights, resulting in discrete layers of foliage attractive to diverse bird species (McArthur and McArthur 1961). Such vertical

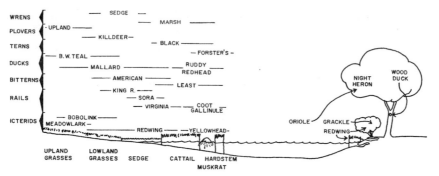

Figure 1.1. Habitat use by various bird families in relation to vegetation structure and water depth typical of a prairie basin wetland (modified from Weller and Spatcher 1965.

layering in herbaceous wetlands includes vegetation that has less actual height change but is of no less important to water-level birds; also it includes physical components like shoreline, vegetation layers, water column depth and structure, and wetland substrate (Fig. 1.1) **Habitat diversity** is a product of various biological and physical features that make multiple sub- or microhabitats available in a small area which are attractive to different species or groups of birds. Here, we will most commonly denote this diversity as **species richness** (i.e., number of species found in an area) and discuss other elements of diversity later.

The habitat concept also infers that this spatial site meets the "psychological" perception by a bird species, termed **umwelt** by German workers to infer the animal's view of its environment as influenced by its special senses; Klopfer (1969) pointed out that this means mostly visual clues in birds. A simple model of some features of habitat recognition by birds was developed to demonstrate the role that habitat plays in separating closely related species such as pipits, in which one species used wet meadows and others select more vegetated and elevated sites (Svardson 1949). These perspectives resulted in the concept of **habitat patterns** as a way of describing features of habitat structure that we humans recognize as important to birds but have difficulty quantifying (e.g., water, open horizon, plant height or density, etc.)(Williams, Russell and Seitz 1977). Such clues to suitable habitat presumably are perceived from the air as well as on-water, in-water, and underwater and consider both visual appearance and feel. One clue to such visual components comes from accidents made by night-time migrants that land on wet streets and are stranded until daylight and even then may have difficulty taking off.

Based on studies of the behavior of young waterbirds, we also expect that some of this response to habitat is **innate** (i.e., instinctive in newly hatched birds) and some may be **learned** in early imprinting to environmental features. For example, some workers have noted that young, captive ducklings imprinted on humans still frequent species-oriented habitat preferences when possible (Fabricius 1951). The influence of learning was suggested to me early in my studies when I found the nest of a banded Redhead Duck that had been hand-reared and released the previous year but returned to nest in flooded forest rather than in the typical open marsh where the egg had been collected and where adults normally breed. Did this forest canopy provide the habitat image acquired during its early life in an incubator and covered rearing pen?

It is obvious from the above that birds make decisions about which habitat to use based on innate and learned behavior and by trial-and-error testing of features and resources. They go through a process of **selection** and ultimately use of a chosen habitat. But habitat **use** may not infer habitat **preference** – unless a bird has choices including the best (i.e., where they are most success-ful) – a most difficult thing to demonstrate unless one knows all of the bird's requirements.

Habitat is a general, descriptive term often used without stated scales; it is enhanced by the addition of temporal scales essential to reflect annual **life-cycle stages,** and by spatial **scales** to denote structural differences or geograph-ical coverage. Considerable effort now is directed toward larger scales termed **landscape** because of current technological advances both in computer-based imagery and in statistical methodology (Forman and Godron 1986). Several spatially separated **habitat units** may be necessary to complete the annual cycle, even in nonmigratory but mobile species, and all contribute to repro-ductive success and population maintenance. However, the habitat structure of these sites may differ markedly, as may food and other resources. Compare, for example, the similarity of wooded feeding and resting sites used by migra-tory North American Wood Ducks during either breeding or wintering with the habitats of Redhead ducks, which may dive for invertebrate foods in densely vegetated prairie potholes during the breeding season but dabble for seagrass rhizomes in hypersaline estuaries during winter.

Regardless of the temperature, probably fewer wetland birds reside year-round in small home ranges than is common to terrestrial species, because sea-sonal changes in wetlands are so dramatic. Water levels rarely remain the same in most wetland types: vegetation dies and decomposes; foods such as seeds, foliage, and invertebrates vary in abundance and distribution; water tempera-tures influence oxygen levels and thereby aquatic animal abundance; and freeze-ups severely affect what birds can do. Often, the changes are simply less conspicuous and reflect distribution of resources on a smaller scale, demand-ing that investigators achieve more precise assessment.

Whereas the term habitat deals with spatial and structural components of the living space, the word **niche** (and especially **ecological niche**) is a concept describing the role of a species in a community of organisms living together. The meaning of this term was broadened by a multidimensional perspective of factors that reflect the environmental requirements of the species (Hutchinson 1967). Despite many attempts to clarify the term, uncertainty and confusion still exist (Begon *et al.* 1996, Brown 1995, Odum 1971, Patten and Auble 1981, Whittaker *et al.* 1973), but that is not uncommon with theoretical concepts. For our purposes, its importance is mainly to denote what the species does and how it lives in the community. When used in reference to food-chains and relationships, **trophic niche** is a useful term (Odum 1971) that facilitates understanding of energy transfer between organisms of the ecosystem (e.g., herbivore, carnivore, or omnivore) and the manner in which it occurs (e.g., predator, grazer).

When dealing with groups of species using the same habitat, the word **assemblage** is a convenient term to infer a group of bird species in one place at one time – without implying any organized interrelationships. Obviously, birds that are together by chance or preference may interact in social feeding, competition for resources, and predation. The term **community** often is used in this way (Begon *et al.* 1996) whereas other researchers think of it more as the product of evolutionary interactions resulting in organized responses as a unit (see review in Wiens 1989). My use of the term community infers no fixed grouping but is convenient when discussing interacting species such as those involved in competition or predator–prey relations, which are expected whenever mixed species are involved. An analysis of the bird community by such strategies and roles is termed **community structure** and should not be confused with habitat or habitat structure.

Compared with the migration of other vertebrates, most birds are not restricted to one wetland to fulfill feeding or other maintenance requirements (except when flightless) but may use several adjacent wetlands (e.g., **wetland complex**) at a landscape scale. Fish are dependent on connectivity between wetlands, and amphibians, reptiles, and smaller mammals expose themselves to predation if they move very far. Massive movements do occur in these groups, associated with mating and breeding sites (amphibians), egg laying (fish, reptiles), dispersal of young (mammals), or drought (all), but at a far smaller scale and in a restricted time frame compared with that seen in birds.

One cannot consider resource-use patterns of waterbirds without attempting to clarify the long-term evolutionary changes from the presumably learned or at least more-adaptable behaviors used in daily functions and needs. Although not in traditional use, I will tend to separate those long-term and genetically based adaptations that affect how birds satisfy life requirements (e.g., food habits, breeding behavior, physiology, selection of geographic range for breeding and wintering) by use of the term **strategy** to infer a complex of steps that accomplishes some major life function for the bird and that presumably has a long-term evolutionary basis (similar to that used by Ellis *et al.* (1976) for food selection). The term **tactic** (or **technique** of DeGraaf, Tilghman and Anderson (1985)) will be used to indicate the more proximate and adaptable methods used to satisfy an immediate short-term objective (e.g., gathering food, roosting behavior, nest sites). Any such separation is always debatable, but the concepts are helpful.

Discussion of bird use of habitats has been much enhanced by the concept of **guild** (Root 1967), a term that describes a group of species that utilize the same resource in a similar manner without reference to taxonomic relationships. This term is used differently by various workers at various scales. One

Figure 1.2. Model of comparative factors and their potential interactions on the reproductive success of birds in a given habitat (from Karr, 1980, with permission).

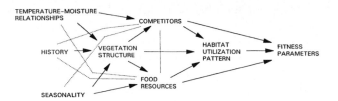

could talk about Muskrats, Common Carp, and swans as a single guild grazing in shallow water on submergent vegetation, but that is perhaps too gross to be useful. More often, one identifies a group of more similar species that exploit a resource such as food or nest sites in a prescribed habitat such as a stand of emergent vegetation. Thus, we can focus on patterns of use based on how they operate (e.g., swimming invertebrate strainers or wading piscivores) rather than their taxonomic relationship (e.g., ducks, herons, sparrows).

Having given the background of essential tools used in the book as well as clarifying terms, I will next introduce the habitat of focus, wetlands. It is important, however, that we do not fall into the trap of ignoring the many other influences on the evolution of the features and species of birds in these habitats and which will influence their success over time. Birds that cannot succeed at maintaining a population do not use a habitat for long, but such deterministic influences may be genetically based morphological and anatomical features of importance and other environmental influences that are rarely discussed in relation to habitat. Several simple models have been used by various workers to place these issues in perspective in the life history of the species and to outline a framework for the analysis of interactions and relative importance of habitat features that determine the presence and success of the species (e.g., Karr 1980, Price and Bock 1983)(Fig. 1.2). But few facets of the life history of obligate wetland birds are not influenced by water and wetland resources.

References

Aldrich, J. W.(1945). Birds of deciduous forest aquatic succession. *Wilson Bulletin* 57, 243–5.

Allen, A. A. (1914). The red-winged blackbird, a study in the ecology of a cat-tail marsh. *Proceedings of the Linnaean Society of New York* 24, 43–128.

Allen, R. P. (1942). *The roseate spoonbill.* Research Report No. 2. New York: National Audubon Society.

Bailey, R. G. (1995). *Descriptions of the ecoregions of the United States,* 2nd edn, *Miscellaneous Publication No. 1391.* Washington, DC: US Forest Service.

Beecher, W. J. (1942). *Nesting birds and the vegetative substrate.* Chicago, IL: Chicago Ornithological Society.

Begon, M., Harper, J. L., and Townsend, C. R. (1996). *Ecology,* 3rd edn. Oxford: Blackwell Science.

Bennett, L. J. (1938). *The blue-winged teal: its ecology and management.* Ames, IA: Collegiate Press.

Bent, A. C. (1919–68). *Life histories of North American birds* (26 vol.) Washington, DC: US National Museum.

Brown, J. H. (1995). *Macroecology*. Chicago, IL: University of Chicago Press.

Burger, J. (1985). Habitat selection in temperate marsh-nesting birds. In *Habitat selection in birds*, ed. M.L. Cody, pp. 253–81. Orlando, FL: Academic Press.

Campbell, B. & Lack, E. (1985). *A dictionary of birds*. London: T & A D Poyser.

Cody, M. L. (1981). Habitat selection in birds: the roles of vegetation structure, competitors, and productivity. *Bioscience* 31, 107–13.

Cody, M. L. (ed.) (1985). *Habitat selection in birds*. Orlando, FL: Academic Press.

DeGraaf, R. M., Tilghman, N. G., and Anderson, S. H. (1985). Foraging guilds of North American birds. *Environmental Management* 9, 493–536.

Ellis, J. E., Weins, J. A., Rodell, C. F., and Anway, J. C. (1976). A conceptual model of diet selection as an ecosystem process. *Journal of Theoretical Biology* 60, 93–108.

Fabricius, E. (1951). Zur Ethologie jungen Anatiden. *Acta Zoologica Fennica* 68, 1–175.

Forman, R. T. T. and Godron, M. (1986). *Landscape ecology*. New York: Wiley.

Gibbons, A. (1997). Did birds fly through the K–T extinction with flying colors? *Science* 275, 1068.

Hall, L. S., Krausman, P. R., and Morrison, M. L. (1997). The habitat concept and a plea for standard terminology. *Wildlife Society Bulletin* 25, 173–82.

Hilden, O. (1965). Habitat selection in birds: a review. *Annales Zoologici Fennici* 2, 53–75.

Hutchinson, G. E. (1967). *A treatise on limnology*, Vol. 2. *Introduction to lake biology and limnoplankton*. New York: Wiley.

James, F. C. (1971). Ordination of habitat relationships among breeding birds. *Wilson Bulletin* 82, 215–36.

Karr, J. R. (1980). History of the habitat concept in birds and measurement of avian habitats. *International Ornithological Congress* 17, 991–7.

Klopfer, P. H. (1969). *Habitats and territories: a study of use of space by animals*. New York: Basic Books.

Lack, D. (1933). Habitat selection in birds. *Journal of Animal Ecology* 2, 239–62.

Lack, D. (1968). *Ecological adaptations for breeding in birds*. London: Methuen.

MacArthur, R. H. and MacArthur, J. W. (1961). On species diversity. *Ecology* 42, 594–8.

Marin, V. H. (1997). General system theory and the ecosystem concept. *Bulletin of the Ecological Society of America* 78, 102–4.

Martin, A. C., Zim, H. S., and Nelson, A. L. (1951). *American wildlife and food plants: a guide to wildlife food habits*. New York: McGraw-Hill.

McAtee, W. L. (1939). *Wildfowl food plants*. Ames, IA: Collegiate Press.

Morrison, M. L., Marcot, B. G., and Mannan, R. W. (1992). *Wildlife–habitat relationships*. Madison, WI: University of Wisconsin Press.

Niemi, G. J. (1985). Patterns of morphological evolution in bird genera of New World and Old World peatlands. *Ecology* 66, 1215–28.

Odum, E. P. (1971). *Fundamentals of ecology*. Philadelphia, PA: Saunders.

Orians, G. H. (1980). *Monographs in Population Biology,* vol. 14: *Some adaptations of marsh-nesting blackbirds*. Princeton, NJ: Princeton University Press.

Orians, G. H. and Willson, M. F. (1964). Interspecific territories of birds. *Ecology* 45, 736–45.

Patten, B. C. and Auble, G. T. (1981). System theory of the ecological niche. *American Naturalist* 117, 893–922.

Price, F. E. and Bock, C. E. (1983). *Studies in avian biology* No. 7; *Population ecology of the Dipper* (Cinclus mexicanus) *in the Front Range of Colorado*. Cooper Ornithological Society.

Rojas, L. M., McNeil, R., and Cabana, T. (1997). Diurnal and nocturnal visual function in the tactile foraging waterbirds: the American white ibis and the black skimmer. *Condor* 99, 191–200.

Root, R. B. (1967). The niche exploitation pattern of the Blue-gray Gnatcatcher. *Ecological Monographs* 37, 317–50.

Schorger, A. W. (1947). The deep diving of the Loon and the Old-squaw and its mechanism. *Wilson Bulletin* 59, 151–9.

Southwood, T. R. E. (1977). Habitat, the template for ecological strategies? *Journal of Animal Ecology* 46, 337–65.

Svardson, G. (1949). Competition and habitat selection in birds. *Oikos* 1, 157–74.

Verner, J. and Willson, M. F. (1966). The influence of habitat on mating systems of North American Passerines. *Ecology* 47, 143–7.

Weller, M. W. and Fredrickson, L. H. (1974). Avian ecology of a managed glacial marsh. *Living Bird* 12, 269–91.

Weller, M. W. and Spatcher, C. E. (1965). *Role of habitat in the distribution and abundance of marsh birds*. Special Report No. 43. Ames, IA: Iowa State University Agriculture and Home Economics Experiment Station.

Whittaker, R. H., Levin, S. A., and Root, R. B. (1973). Niche, habitat, and ecotope. *American Naturalist* 107, 321–38.

Wiens, J. A. (1989). *The ecology of bird communities*, vols 1 and 2. Cambridge: Cambridge University Press.

Williams, G. L., Russell, K. R., and Seitz, W. (1977). Pattern recognition as a tool in the ecological analysis of habitat. In *Classification inventory, and analysis of fish and wildlife habitat: proceedings of a national symposium*, Phoenix, AZ, pp. 521–31. Washington, DC: US Fish and Wildlife Service, Office of Biological Services.

Willson, M. F. (1966). The breeding ecology of the Yellow-headed Blackbird. *Ecological Monograph* 36, 51–77.

2

Wetlands: what, where, and why

An explanation of avian habitat use and distribution in wetlands involves understanding what wetlands are, identifying the geological setting (**geomorphology**), water sources (**hydrology**), and climatic influences of the site that explain why and where these unique biotic communities occur. In addition, any explanation must describe the characteristic biological and structural diversity that birds may perceive. To facilitate discussion and identification of essential features of various wetland types, I must review some current wetland classification systems used in mapping and regulatory decisions because this is the context in which many readers deal, but I shall select those especially useful in descriptions of bird habitats. For legal purposes, it is often necessary to identify the boundary between a wetland and the adjoining upland or the open water of more aquatic habitats, but this discussion will focus on features that birds seem to perceive and on boundaries that appear to influence them.

2.1 Classification of wetlands

Wetlands are not easily defined because they range from near-terrestrial to the aquatic, and because many are dynamic in water regime and, therefore, variable in vegetation patterns and bird use. Some vary rather predictably and dramatically by season or year, and many also are subject to long-term variation owing to large-scale climate patterns and cycles. Therefore, wetlands are sometimes not recognizable by standing water, and their vegetation can be equally deceiving. In addition to the presence of shallow water or wet soil periodically, two other key features of wetlands are the periodic presence of water-adapted plants (**hydrophytes**), which range from mosses to giant trees, and **hydric soil,** with biochemical features influenced by anaerobic conditions of flooding (Cowardin *et al.* 1979).

Our interest here is mainly in wetlands as biological communities, for which the Cowardin *et al.* (1979) classification is mandated in regulatory situations in the United States, and is also now popular worldwide. It involves five **systems** (not to be confused with ecosystem) delimited in most cases by uplands on the shallow side and by prescribed water depth (2m), source of salinity (coastal or inland), and vegetation patterns in the deeper portion. The five systems are **Marine** (shallow coastal saltwater), **Estuarine** (brackish coastal water), **Lacustrine** (relatively shallow, open, freshwater lakes or their sparsely vegetated margins), **Palustrine** (marshy fresh or *inland saline* waters or the more dense shoreward vegetated margins of larger water bodies), and **Riverine** (used in this system to mean those wetlands found only within the river channel, as opposed to river-formed wetlands). Hierarchical subdivisions within each system incorporate terms that help describe the habitat structure or vegetation of the wetland type (Table. 2.1). For example, a Palustrine wetland could be dominated by moss–lichens (as in a bog), by herbaceous, emergent vegetation (e.g., cattail marsh), by shrubs or scrub (shrub swamp), or by water-tolerant trees (wooded or forested swamp or bottomland forests). Examples of some of these systems and classes are evident in photos throughout the book particularly in those of Plate I. These dominant vegetation types are products of several factors: the presence of water over time (**hydroperiod**), **water depth, seasonality**, temperature regimes in the locale, and other factors that will become obvious later. But in all cases, they form diverse habitats attractive to different bird and other animal species.

The Canadian wetland classification system is more strongly influenced by cold-climate bog wetlands but uses commonly recognized terms such as bog, fen, swamp, marsh, and shallow water (National Wetland Working Group 1988). Regional denotations such as arctic, coastal, and similar terms help to focus on the extensive boreal areas so dominated by wetlands. Similar interests are evident in several Scandinavian countries, where limnological terms are favored that reflect long-term succession from deep, clear lakes poor in nutrients (**oligotrophic**) to the more shallow, vegetation-and nutrient-rich lakes of marshes (**eutrophic**)(e.g., Kauppinen and Vaisanen 1993). Several of these publications trace the history of the development of various classification systems devised, and the terms are many. In contrast, the Cowardin system was designed to eliminate earlier and often conflicting terms and create a hierarchical approach that allows any subcategories which may be necessary for the description and differentiation of wetlands.

The Cowardin system considers as wetlands areas that are not in a discrete basin but which are wet long enough to induce wetland plant or animal response. While most wetlands are basins that capture water from precipitation, flooding, or underground aquifers, there are flat areas that hold water for a long time or hillside slopes in hilly regions or where underground water percolation reaches plant root systems – sometimes without even appearing wet on the surface. Thus, water and the geomorphic setting drive the origin of wetland communities, and these causal and formative processes are best elaborated in the **Hydrogeomorphic Method** (HGM)(Brinson 1993). There are five categories: **depression** or **basin** wetlands, which may receive water from surface runoff or groundwater; **fringe** wetlands, which develop along large

Wetlands: what, where, and why

System	Subsystem	Classes
Marine	Subtidal	Rock
	Intertidal	Unconsolidated
		Aquatic bed
		Reef
Estuarine	Subtidal	Rock
	Intertidal	Unconsolidated
		Aquatic bed
		Reef
		Emergents
		Shrub–scrub
		Forested
Riverine	Tidal	Rock
	Lower perennial	Unconsolidated
	Upper perennial	Aquatic bed
	Intermittent	Emergent
		Streambed
Lacustrine	Limnetic	Rock
	Littoral	Unconsolidated
		Aquatic bed
		Emergent
Palustrine	(None)	Rock
		Unconsolidated
		Aquatic bed
		Moss-lichen
		Emergent
		Shrub–scrub
		Forested

Source: From Cowardin *et al.* (1979)

lakes or oceans; and **riverine** wetlands (typically linear), which are a product of running water (differing from the Cowardin System, which includes only the stream channel). Less well-known are groundwater **slope** wetlands, which have wet soils or actual discharge areas where the water induces hydrophyte growth even on hillsides, and **flatlands,** where high rainfall runs off very slowly and is lost by permeating the substrate but also by evaporation and plant transpiration.

For clarification, ponds and lakes are common and often poorly defined terms for more open water-bodies with abrupt banks that may have little or no emergent vegetation. Cowardin *et al.* (1979) separated ponds from lakes on the basis of size, using 8ha (20 acres) as the breaking point. The term pond is often used for small, water-bodies with abrupt shorelines, deep basins, and open water (only for constructed water areas in some regions). Lakes typically differ from wetlands in having a higher percentage of open water, little emergent vegetation, and well-defined shorelines produced by water- or ice-formed boundaries or banks.

The boundary between a wetland and the upland is important ecologically because it forms a transition zone of varying water regimes and plant communities. At this point, various microhabitats are created and selected for by wetland birds and by those species that tap resources of both upland and wetland vegetation (Fig. 1.1). In wetlands with dynamic water regimes, edges often change seasonally and annually, and birds favoring that edge select habitat upslope or downslope to follow features or resources important to them. Boundaries are of concern in a conservation and regulatory sense because they influence decisions on land-use, and there are well-described techniques for delineation that involve ecological measures of plant species richness and composition change over the edge gradient (Federal Interagency Committee for Wetland Delineation 1989).

2.2 Factors influencing wetland formation

One would expect to find more wetlands where there is higher precipitation, but the presence of wetlands is influenced by the landforms or soil structure that traps the moisture. Thus, the tens-of-thousands of small wetlands, ponds, and lakes in the Prairie Pothole Region of the northern USA and southern Canada result more from the undulating landform left by glaciers than to the amount of rainfall. Because of various basin depths, their number and size vary in response to the precipitation cycle, influenced especially by winter snow drifts and spring rains. In regions where topography creates larger and deeper lake (rather than wetland) basins, marsh-like (Palustrine) vegetation forms on the shoreward fringes and open-water characteristics (Lacustrine) on the lakeward side. Mountains in the eastern USA influence drainage onto the low-lying coastal plain, resulting in fewer but larger and more permanent swamps (Wharton *et al.* 1982). In areas of lower rainfall and high evaporation, as in the western USA and other arid plains and basins of the world, wetlands may be very large but vary in size with the amount of snow accumulated in high mountains that melts in that season. Lakes tend to form in valleys between mountain uplifts until drainage develops, and these may form closed basins where water pools and evaporates (sumps and pans). In North America's Great Basin, the Great Salt Lake, Malheur National Wildlife Refuge in Oregon (Duebbert 1969), the Stillwater National Wildlife Refuge, and the Carson Sink in Nevada all fall into this category, and the Cheyenne Bottoms in Kansas has similar characteristics (Zimmerman 1990). Some of the rift lakes of East Africa show long-term water cycles that dramatically affect bird use. The water dynamics of the Great Salt Lake (Kadlec 1984) is a dramatic example, and one of my favorite study areas there changed over a 30-year period from a near-fresh marsh to salt lake bottom through long-term changes in precipitation patterns. It is only a matter of time until the area is "recycled" by drying to become first extremely saline and eventually to be freshened by stream flow to more characteristic brackish or fresh marsh.

River forces are major creators of wetland through scouring currents and meanders and cutoffs that form oxbow wetlands and lakes. Riparian or bank-side wetlands are induced by bank sediment deposition, forming retainimg walls behind which backwater swamps form in the low areas. Valleys are broader near the sea, current is slower except in flood, and meanders are greater. There are several river classification systems in use; the Cowardin system uses a common standard: Lower Perennial for broad river channel areas near the sea as opposed to Higher Perennial or Intermittent for the narrow highland streams or other channels with temporary flows. Some examples from other countries demonstrate the importance of riverine wet-lands to society, and the physiographic influences on water regimes. In Africa, riverine runoff along short channels into low basins and flats (e.g., Etosha Panne in Namibia and Okovango Swamp in Botswana) are products of erratic rainfall along several contributing drainages. Wildlife and local human populations have been geared to capitalizing on the resultant resources for thousands of years.

Where rivers meet the sea (Estuarine), outpouring of sands and silts form major deltaic marshes. Much of coastal Louisiana falls in this category from the major flows of the Missouri, Illinois, Ohio, and many other streams that enter and form the mighty Mississippi River (Gosselink 1984). It is also our modification of this flow via dams that is now robbing the delta of the life-giving silt upon which plants can grow. As a result of this and other changes induced by humans, marsh loss in Louisiana is estimated at about 40 square miles per year!

River deltas also may create extensive tidal but freshwater marshes by virtue of tidal extremes and backup of inflow (Odum *et al.* 1984). The famous "people of the marsh" (Maxwell 1957) who once lived isolated lives in the Tigris–Euphrates delta of Iraq/Iran depended on products of the delta and remote water sources, and recent diversions have seriously affected their way of life.

Other wetlands along the coast are also an ultimate product of this river-based clay, silt, and sand after it is moved seaward and redistributed along the coast in the direction of the prevailing winds and currents. Most directly, silt/sand depositions forms spits and barrier islands that create bays, estuaries, and lagoons. This same patterns closes passes and channels made by other forces such as hurricanes and severe winter storms (Chabreck 1988). Over time, and often with sea-level change and climate change, particles may be deposited as beaches. Old beach rows and declining seas levels result in what in western Louisiana and extreme eastern Texas are called Chenier Marsh, named after the wooded ridges or "cheniers" that really are ancient beach ridges; these parallel barriers result in a series of fringing marshes that become more shallow and fresher in more shoreward areas. Still more ancient river out-pourings resulted in ancient depositions that subsequently are moved far inland by coastal winds, forming sand/silt dune and swale areas, so that fresh-water ponds and lakes may be common even along rather dry coastlines.

Such beach/ridges structures also are major influences on wetland forma-tion at various scales along large lakes such as the Great Lakes of North America (Prince and D'Itri 1985). In the bordering states and provinces, much of which was once a part of post-glacial Lake Agassiz, lake decline left ancient beaches that major highways cross without most drivers being aware of their existence, history, or formation, and major wetland areas between these beaches (Glaser 1987).

Marine coasts may demonstrate different profiles dependent upon shore-line materials and gradient (e.g., cliff versus mudflat), which affect how long tidewater remains. Shoreline mudflats and sandbars are especially crucial to shorebirds, gulls, and terns. Substrate characteristics influence how long water is retained in the sand, silt, or mud, and, therefore, how suitable it is for prey such as the clams and worms that are so important to birds. These features of coastal wetlands strongly influence bird species composition, bird density, and timing of foraging and roosting/resting activities.

A unique wetland region, the coastal tundra of the Arctic, shows the impor-tance of physical influences on wetland development. These open but shallow wetlands are formed on permafrost by solar melting of the substrate wherever water collects and gathers heat (Bergman et al. 1977). Once formed, they tend to enlarge both by melting of the substrate and by wind erosion. But these wet-lands can be huge but very shallow because of the vertical temperature equilib-rium between the water (in summer) and the underwater ice base that remains all year. With precipitation of less than 38cm per year, water in the shallow basins is derived mostly from snowfall driven by winds. Once in the pools, evaporation results in drying so few plants grow in the water, but peat accumulation forms substrates rich in invertebrates, and thereby attractive to skimming and probing birds.

The geological processes that form wetlands and lakes also include tectonic forces such as major earth-shaping forces; earthquakes and volcanic action are localized but dramatic examples. Several large inland cypress swamps (Reelfoot Lake of northwestern Tennessee and Caddo Lake of eastern Texas and western Louisiana) may have been formed by earthquakes as recently as the 1800s. Earthquakes have modified extensive wetland areas in the Copper

River delta of Alaska. Lakes with associated wetlands areas have been formed by volcanic actions in the Galapagos Islands and lost and reborn in the Mount St. Helens earthquake in southern Washington.

Biological forces also influence and, in some cases, actually form locally abundant and important wetlands. The most common animal that forms wetlands is the beaver, which is widespread over North America and can have major impacts on small stream systems. They are unpopular because they kill trees through flooding and cutting, but this is part of a process of succession that produces other wetland plant species in the resulting flooded areas. They are regarded as major stream-flow regulators by some hydraulic engineers. Alligators in southern marshes create openings in herbaceous vegetation and dig holes and burrows that strongly influence the presence of open water and the potential survival of fish, snakes and amphibians, and insects in times of drought (Kushlan 1976).

Plant production itself provides a water-holding mechanism through the development of wet, organic soils. Bogs "grow" on water-holding substrates, sometimes forming islands with surrounding moats of water and often forming "patterned ground" where slope and peat are involved (Damman and French 1987, Glaser 1987). A similar but more dynamic process is the development of floating islands of grass mats, common at flooding stages of herbaceous marshes or swamps, tree islands, which also include herbaceous plants (e.g., the Florida Everglades), and tropical mangroves, which may establish on such islands (Odum, McIvor, and Smith 1982).

In these ways, wetland habitats are products of many physical and biological processes and events at various scales, and they are continuous. It is mainly our short-term and often static view that leaves an impression of wetlands as fixed entities. The impact of human development has often changed many of the influential parameters, and especially the rates at which such changes take place.

Once wetlands have formed, many factors influence the integrity, resilience, and existence of these biotic communities; their productivity influences which regional bird species use them, in which season of the year, and for how long. We will discuss a number of examples later, but it will be helpful to generalize the water regimes typical of the wetland types mentioned above because these patterns and mean depths are reflected in the responses of plants that influence bird use (Fig. 2.1). These are hypothetical estimates of the average ranges of water fluctuation in various wetland types in a season. At any one time, water levels could be at one extreme or the other, but it is the average depth and seasonal timing of this water that dictates the characteristic vegetation of the site. Therefore, lakes are lakes because they tend to be wet most of the time and not conducive to the establishment of emergent vegetation. Generally, lakes tend to be more stable than constructed reservoirs but both tend to be deeper than 2m and to lack emergent vegetation (Cowardin *et al.* 1979). Floodplain wetlands are characterized by extreme water-level

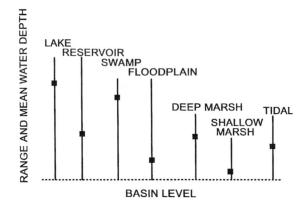

Figure 2.1. Typical patterns of water level in several common wetland types. Ranges and means vary by region but reflect relative stability (modified from Weller 1994).

fluctuations, resulting from seasonal river flows. Newly created bars and shallow shorelines are frequented by annual plants, animals with short life-cycles, and more adaptable birds geared to feeding under those shallow conditions. Shallow but semipermanently flooded wetland basins are rich in more persistent and perennial emergents and even in aquatic plants, and, as a result, with birds favoring deeper water. Long-lived woody plants more often develop in backwater basins that capture and hold water for part of the year. These water dynamics and substrate disturbance patterns should not be construed as reducing productivity of the system, which often is greatest in seasonally dynamic wetlands. Tidal systems could be illustrated in any magnitude of extremes, but in (Fig. 2.1) a median point is shown with a range from tidal highs to lows. This range differs by latitude and geomorphic settings from the extremes observed in the Bay of Fundy and Tierra del Fuego (over 60 feet, 18 m) to a modest 12–18 inches (30–45cm) in the Gulf of Mexico. These ranges influence vegetation development and bird use because they require adaptations to extremes and daily and seasonal timing. At any site, the celestial influences are predictable but wind tides are an especially important influence on water depths in lesser tide ranges.

To understand how these general wetland characteristics influence avian habitat selection, we will next examine life-history characteristics and adaptations of the birds using these diverse and often variable habitats.

References

Bergman, R. D., Howard, R. L., Abraham, K. F., & Weller, M. W. (1977). *Waterbirds and their wetland resources in relation to oil development at Storkerson Point, Alaska.* Resource Publication 129. Washington, DC: US Fish & Wildlife Service.

Brinson, M. M. (1993). *A hydrogeomorphic classification of wetlands.* Wetlands Research Program Technical Report WRP-DE-4. Vicksburg, MS: US Army Corps of Engineers.

Chabreck, R. H. (1988). *Coastal marshes: ecology and wildlife management.* Minneapolis, MN: University of Minnesota Press.

Cowardin, L. M., Carter, V., Golet, F. C., & LaRoe, E. T. (1979). *Classification of wetlands and deepwater habitats of the United States.* FWS/OBS 79/31. Washington, DC: US Fish & Wildlife Service, Office of Biological Services.

Damman, A. W. H. & French, T. W. (1987). *The ecology of peat bogs of the glaciated Northeastern United States: a community profile.* Biological Report 85(7.16). Washington, DC: US Fish & Wildlife Service.

Duebbert, H. E. (1969). *The ecology of Malheur Lake.* Refuge Leaflet No. 12. Washington, DC: US Fish & Wildlife Service.

Federal Interagency Committee for Wetland Delineation (1989). *Federal manual for identifying jurisdictional wetlands.* Cooperative Technical Publication. Washington DC: US Army Corps of Engineers, US Environmental Protection Agency, US Fish & Wildlife Service, & US Soil Conservation Service.

Glaser, P. H. (1987). *The ecology of patterned boreal peatlands of northern Minnesota: a community profile.* Biological Report 85(7.14). Washington, DC: US Fish & Wildlife Service.

Gosselink, J. G. (1984). *The ecology of delta marshes of coastal Louisiana: a community profile.* FWS/OBS-84/09. Washington, DC: US Fish & Wildlife Service.

Kadlec, J. A. (1984). Rising Great Salt Lake inundates marshes. *National Wetlands Newsletter* 6, 2–3.

Kauppinen, J. and Vaisanen A. (1993). Ordination and classification of waterfowl communities in south boreal lakes. *Finish Game Research* 48, 3–23.

Kushlan, J. A. (1976). Wading bird predation in a seasonally fluctuating pond. *Auk* 93, 464–76.

Maxwell, G. (1957). *The people of the reeds.* New York: Pyramid Books.

National Wetland Working Group (1988). *Wetlands of Canada.* Montreal, Quebec: Polyscience.

Odum, W. E., McIvor, C. C., & Smith III, T. J. (1982). *The ecology of the mangroves of South Florida: a community profile.* FWS/OBS- 81/24. Washington, DC: US Fish & Wildlife Service.

Odum, W. E., Smith III, T. J., Hoover, J. K., & McIvor, C. C. (1984). *The ecology of tidal freshwater marshes of the United States east coast: a community profile.* FWS/OBS- 83/17. Washington, DC: US Fish & Wildlife Service.

Prince, H. H. & D'Itri, F. M. (1985). *Coastal wetlands.* Chelsea, MI: Lewis.

Weller M. W. (1994). *Freshwater marshes: ecology and wildlife: management.* 3rd edn. Minneapolis, MN: University of Minnesota Press.

Wharton, C. H., Kichens, W. M., Pendleton, E. C., & Sipe, T. W. (1982). *The ecology of bottomland hardwood swamps of the Southeast: a community profile.* FWS/OBS- 81/37. Washington, DC: US Fish & Wildlife Service, Biological Services Program.

Zimmerman, J. L. (1990). *Cheyenne Bottoms, wetland in jeopardy.* Lawrence, KS: University Press of Kansas.

Further reading

Cowardin, L. M. & Johnson, D. H. (1973). *A preliminary classification of wetland plant communities in north-central Minnesota.* Special Scientific Report-Wildlife 168. Washington, DC: US Fish & Wildlife Service.

Fredrickson, L. H. & Taylor, T. S. (1982). *Management of seasonally flooded impound-*

ments for wildlife. Resource Publication No. 148. Washington, DC: U S Fish & Wildlife Service.

Hammer, D.A. (1992). *Creating freshwater wetlands*. Chelsea, MI: Lewis.

Harris, S. W. & Marshall, W. H. (1963). Ecology of water-level manipulations of a northern marsh. *Ecology* 44, 331–42.

Kantrud, H. A., Krapu, G. L., & Swanson, G. A. (1989). *Prairie basin wetlands of the Dakotas: a community profile*. Biological Report 85(7.28). Washington, DC: US Fish & Wildlife Service.

Martin, A. C., Hotchkiss, N., Uhler, F. M., & Bourn, W. S. (1953). *Classification of wetlands of the United States*. Special Scientific Report-Wildlife No. 20. Washington, DC: U S Fish & Wildlife Service.

McComb, A. J. & Lake, P. S. (eds.) (1988). *The conservation of Australian wetlands*. Chipping Norton, NSW: Surrey Beatty & Sons.

Mitsch, W. J. and Gosselink, J. G. (1993). *Wetlands*, 2nd edn. New York: Van Nostrand Reinholdt.

Niering, W. A. (1985). *Wetlands*. Audubon Society Nature Study Guides. New York: Alfred A. Knopf.

Reed, P.B. Jr.(1988). *National list of plant species that occur in wetlands: national summary*. Biological Report 88(24). Washington DC: US Fish & Wildlife Service.

Reid, G. K. (1961). *Ecology of inland waters and estuaries*. New York: Van Nostrand.

Riemer, D. N. (1984). *Introduction to freshwater vegetation*. Westport, CT: AVI Publishing.

Scott, D. A. and Carbonell, M. (1986). *A directory of Neotropical Wetlands*. Slimbridge, UK: International Waterfowl Research Bureau and International Union for Conservation of Nature and Natural Resources.

Sculthorpe, C. D. (1967). *The biology of aquatic vascular plants*. London: Edward Arnold.

Stewart, R. E. and Kantrud, H. A. (1971). *Classification of natural ponds and lakes in the glaciated prairie region*. Research Publication No. 92. Washington, DC: US Fish & Wildlife Service.

Stewart, R. E. & Kantrud, H. A. (1972). *Vegetation of prairie potholes, North Dakota, in relation to quality of water and other environmental factors*. Professional Paper 585–D. Washington, DC: US Geological Survey.

Tiner, W. W. Jr (1984). *Wetlands of the United States: current status and recent trends*. Washington, DC: US Fish & Wildlife Service, National Wetlands Inventory.

van der Valk, A. G. (1981). Succession in wetlands: a Gleasonian approach. *Ecology* 62, 688–96.

van der Valk, A. G., & Davis, C. B. (1978). The role of seed banks in the vegetation dynamics of prairie glacial marshes. *Ecology* 59, 322–35.

Weller, M. W. (1978). Management of freshwater marshes for wildlife. In *Freshwater wetlands, ecological processes and management potential*, ed. R. E. Good, D. F. Whigham, & R. L. Simpson, pp. 267–84. New York: Academic Press.

Zieman, J. C. & Zieman, R. T. (1989). *The ecology of the seagrass meadows of the west coast of Florida: a community profile*. Biology Report 85(7.25). Washington, DC: US Fish & Wildlife Service.

3

Major groups of birds that use wetlands

Many books have been written about the taxonomic groups of birds that frequent shallow water or the water's edge, and readers who do not know these birds should use a field guide in conjunction with this text. However, to help the nonspecialist, I have tried to list the major groups worldwide that are obligate waterbirds and I have included other orders less dedicated to water but with examples of wetland representatives (Table 3.1). To appreciate this diversity, it may be helpful to provide a framework for bird distribution at various scales.

3.1 Distribution and adaptations

Bird distribution often is discussed and compared from a geographic perspective (**zoogeography**). At one time, emphasis was mostly on a scale of continents and hemispheres rather than regional habitat, but the importance of vegetation zonation and climate has now become a regular factor in descriptions of groups and subgroups of species. At a somewhat smaller scale, identifying where a particular species is found using geographic and large-scale biotic zones is valuable. This is known as the **geographic range,** which may include geographically separated areas for breeding, wintering, or other needs. A more functional descriptor is **home range,** which denotes an area and habitat features that delimit the living space of an individual, pair, or other specified group. All of the area in which an individual, species, or taxon may be seen (i.e. distribution) is not necessarily habitat (i.e., the area used by the bird), as the bird may pass through less-than-suitable areas to locate the suitable ones.

There is a wide range of adaptation to wetlands among species of any of these taxa. For example a duck swims and strains food from the water with its specialized bill, and most either spend little time in upland areas or at least return to water to roost overnight. Because such birds have difficulty surviving without water, we can term them obligate wetland birds. But there is considerable variation among taxonomic groups; some are less specialized for water (e.g., some storks and cranes) and spend much of their time in dry upland fields, but their morphological adaptations of bill, feet, and legs show long

Table 3.1 *Major taxonomic orders and families or other lower taxa of birds that regularly use wetlands, with general breeding distribution. Examples are given where only a few species of the group are obligate or regular users*[a]

No. species[b]	General breeding distribution
Gaviiformes[c]	
Divers (or loons) (Gaviidae) 5	Arctic and north temperate
Podicipediformes[c]	
Grebes (Podicipedidae) 21	Worldwide
Pelecaniformes[c]	
Frigatebirds (Fregatidae) 4–5	Tropical oceanic
Cormorants (Phalacrocoracidae) 38	
Cormorants (Phalacrocoracinae)	Worldwide but strongly tropical
Shags (Leucocorboninae)	Mostly high latitude and cold waters
Anhinga and darters (Anhingidae) 4	Circumtropical
Pelicans (Pelecanidae) 7–8	Worldwide tropical and temperate
Anseriformes[c]	
Screamers (Anhimidae) 3	South America
Magpie Goose (Anseranatidae) 1	Australia
Ducks, geese and swans (Anatidae) 151–154	Worldwide
Whistling Ducks 8	Circumtropical and warm temperate
White-backed Duck 1	Africa
Geese 16	Arctic and cold temperate; 1 Australia
Swans 8	Arctic and temperate; 1 South America, 1 Australia
Freckled Duck 1	Australia
Spur-winged Goose 1	Africa
Shelducks, sheldgeese, and allies 17–19	Eurasia, Africa, Australia, South America
Steamer-ducks 4	Patagonia (3 of 4 are flightless coastal species)
Dabbling and Wood Ducks 51	Worldwide
Bay Ducks or Pochards 17	Worldwide
Seaducks 20	Arctic and northern cold temperate; 1 extinct Auckland Island Merganser
Stifftail Ducks 9	Worldwide in temperate and tropical regions
Phoenicopteriformes[c]	
Flamingos (Phoenicopteridae) 4–5	Circumtropical
Ciconiiformes[c]	
Herons, egrets, and bitterns (Ardeidae) 65	Worldwide
Whalehead or Shoebill Stork (Balaenicipitidae) 1	Africa
Hammerkop or Hammerhead Stork (Scopidae) 1	Africa
Ibises and spoonbills (Threskiornithidae) 34	Tropical and warm temperate

Table 3.1 *(cont.)*

No. species[b]	General breeding distribution
Storks (Ciconiidae) 19	Temperate and tropics

Falconiformes

Kites, eagles, harriers, hawks (Accipitridae) 239	Worldwide
Osprey[c] 1	Worldwide along coasts, and large rivers and lakes
Harriers[c] 17	Worldwide
Kites, e.g., Snail, Slender-billed, Brahminy, Swallow-tailed	Mostly tropical and warm temperate
Hawks, e.g., Red-shouldered, Black-collared, Common Black-Hawk	Americas (only)
Eagles, e.g., Bald Eagle, Steller's Sea-eagle, African Fishing Eagle	Mostly Northern Hemisphere
Falcons and Caracaras (Falconidae) 63	
Peregrin Falcon	Worldwide
Chimango Caracara	South America

Gruiformes[c]

Rails, crakes, gallinules and coots (Rallidae) 142	Worldwide
Finfoots or sungrebes (Heliornithidae) 3	South America, India, and Africa
Sunbittern (Eurypygidae) 1	South and Central America
Cranes (Gruidae) 14–15	All continents except South America and Antarctica
Limpkin (Aramidae) 1	South and Central America

Charadriiformes[c]

Jacanas (Jacanidae) 8	Central and South America, Africa, Asia, and Australia
Painted-snipe (Rostratulidae) 2	South America, Africa, Asia, and Australia
Sandpipers, phalaropes, and snipes (Scolopacidae)	
Sandpipers, snipe, woodcock (Scolopacinae) 85	Worldwide
Phalaropes (Phalaropinae) 3	Northern hemisphere
Thick-knees (Burhinidae), e.g., Water Thick-knee, Beach Stone-Curlew 9	Worldwide except North America
Oystercatchers (Haematopodidae) 10–11	Worldwide
Stilts and avocets (Recurvirostridae) 10	Worldwide
Crab-plover (Dromadidae) 1	East Africa coast
Coursers and pratincoles (Glareolidae) 17	Europe, Asia
Plovers and lapwings (Charadriidae) 63–66	Worldwide
Lapwings (Vanellinae)	
Plovers (Charadrinae)	
Skimmers (Rynchopidae) 3	Americas, Africa, India, and southeast Asia
Gulls and terns (Laridae)	Worldwide
Skuas and jaegers (Stercorariidae) 7–8	High latitudes, coastal

Table 3.1 *(cont.)*

No. species[b]	General breeding distribution
Gulls (Larinae) 50	Worldwide
Terns (Sterninae) 44	Worldwide

Cucuculiformes

Mangrove and Swamp (or Pheasant) Coucal 2	Tropical and warm temperate
Hoatzin 1	South America

Strigiformes

Barn owl (Tytonidae) 17	Worldwide
African Grass Owl	Only wetland barn owl
Typical owls (Strigidae) 161	Worldwide
Fishing-Owls	Africa and Asia
Barred Owl	North America
Short-eared Owl	Worldwide apart from Africa (where replaced by African Marsh Owl)

Coraciiformes

Kingfishers[c] (Alcedinidae) 87–94	Worldwide in tropical and temperate

Piciformes

Ivory-billed Woodpecker 1 (presumed extinct)	Southern USA and Cuba

Passeriformes (opportunistic wetland use; examples given)

Tyrant Flycatchers (Tyrannidae: Willow, Alder, Kiskadees, Many-colored Rush-Tyrant, Water-Tyrant, White-headed Marsh-Tyrant)	
Ovenbirds (Furnariidae): Wren-like Rushbird, Reedhaunters, Streamcreeper	
Australian warblers (Acanthizidae): Fairy Warbler, Mangrove Warbler	Australia
Crows (Corvidae): Fish Crow, Northwestern Crow	
Vireos (Vireonidae): White-eyed, Mangrove	
Dipper[c] (Cinclidae): 5	Worldwide
Thrushes (Turdidae): Veery, Swainson's Thrush	
Old World Flycatchers (Muscicapidae): Swamp Alseonax or Flycatcher	Africa
Starlings (Sturnidae): Slender-billed Starling	
Babblers (Muscicapidae): Marsh, Chestnut-capped	
Wrens (Troglodytidae): Sedge, Marsh	North America
Titmice (Paridae): Marsh Tit	Europe

Table 3.1 *(cont.)*

No. species[b]	General breeding distribution
Swallows (Hirundinidae): Tree and Mangrove Swallows (many nest near and feed and roost over water)	Worldwide
Old World Warblers (Sylviidae): Arctic, Reed, Marsh, Swamp, Aquatic, Sedge, Paddyfield, Carruther's, Cisticola, Little Rush Warbler	Eurasian, Africa, Australia
Pipits and wagtails (Motacillidae)	
Pipits , e.g., American, Water, Rock	Northern Hemisphere; South Georgia (Antarctic)
Wagtails, e.g., Yellow, Grey, Mountain	Eurasia
Cape Wagtail	Africa (only)
Bush-Shrikes (Malaconotidae), e.g., Marsh Tchagra, Papyrus Gonelek	Africa (only)
Old World Weavers (Plocidae), e.g., Red Bishop, Fan-tailed Widowbird, Marsh Widowbird	Africa (only)
Seedeaters (Fringillidae), e.g., Papyrus Canary	Africa (only)
Sparrows (Emberizidae), e.g., Swamp, Seaside, Nelson's Sharp-tailed, Le Conte's	North America (only)
New World Warblers (Parulidae), e.g., Prothonotary, Bachman's (extinct?), Swainson's, Yellow (incl. Mangrove), Hooded, Yellow-throated, C. Yellowthroat, Waterthrushes	North America (only)
New World blackbirds (Icteridae), e.g., Red-winged, Tricolored, Yellow-headed, Scarlet-headed, Yellow-winged, Brown-and-Yellow Marshbird, Brewer's, Boat-tailed and Great-tailed Grackles	Americas (only)

Notes:

[a] Gill (1995) was used as a basis for taxonomic ordering except for loons.

[b] Number of species varies because of various interpretations in group treatments (see Appendix 1 for scientific names).

[c] A majority of species in the group are obligate wetland users.

adaptation to wet areas and most nest in association with freshwater. The various species of rail demonstrate a range from swimmers and divers (coots) to shoreline or upland-edge nesters (King Rail), but all frequent wetlands regularly and are regarded as obligate users of these resources. Many other orders and families have few obligate wetland bird species, most often reflected in nestsites or food specializations. In the Americas, many blackbirds (family Icteridae) build nests in or near wetlands and rear their young there. Some species are **obligate** (Yellow-headed Blackbird) and others range from the very wet areas to the dry upland (Red-winged Blackbird) (**facultative**); one typically nests in the uplands near a wetland but feeds its young mainly on aquatic insects (Brewer's Blackbird). Other blackbirds are terrestrial or upland but may nest near water or wherever food and cover are suitable (**opportunistic**). In Eurasia, Old-World warblers seem more likely to use wetland vegetation (there are no Icterids or marsh-nesting wrens there) than do New World warblers. However, several New World warblers favor wet bottomland woods where they nest in tree holes or in cane and other grasses of the understory or small gaps in the forest.

Shorebirds show similar segregation by species, although it perhaps is less conspicuous in migration and wintering areas where most of us see them. Their habitat focus tends to be on mudflats and other shoreline vegetation, with species' use of a microhabitat influenced by the amount of water and vegetation as well as by food distribution. Some shorebird species of the drier areas (Killdeer and Grey or Black-bellied Plover) are adaptable to drier sites as well but usually occur near water; yet we always associate the birds with water and consider them at least facultative.

Many species use woody vegetation of the shallows or water's edge, like shrubby willows and alders or trees like mangroves; often these same species use plants of similar lifeform in drier areas. Such habitat use by many birds is more opportunistic; almost all species of swallows use wetlands as feeding and roosting areas, and some (Tree Swallows) nest in holes in snags over water. Kingbirds and some woodpeckers may also nest in similar sites, but it is more an opportunity than a necessity.

Diurnal predators like harriers and Short-eared Owls are even less clearly tied to water; they lack specific adaptations to water but frequent wet areas such as marshes, wet prairie, or flooded meadows, perhaps because of food availability, and they commonly nest there. However, they are not restricted to wetlands and may nest in upland prairie. Still more extreme opportunism is shown by the Red-billed Oxpecker, which picks leeches off a hippopotamus in water but also feeds on ticks on land-based ungulates and nests in upland tree holes and rock crevices.

Some upland birds use wetlands only when they are not wet, and are not considered wetland birds at all: a meadowlark, dove, or Ring-necked Pheasant may nest in a dry wetland and feed there. Pheasants also use wetland cover in winter when the water is frozen, mainly because of the thermal cover the plants or trapped snow provide, but the pheasant has no adaptations for this specific plant community.

3.2 Wetland bird groups

The following synopses outline relevant characteristics of bird groups commonly associated with wetlands worldwide. Although I emphasize habitat adaptations or resource needs, the groups are arranged in a format common to ornithology texts (Gill 1995) and most field guides. The number of species in a taxon may vary significantly with the authority and timing of such taxonomic reviews, some of which are listed among the references. Common English names and scientific names are given in Appendix 1. Later, these birds will be regrouped into community-oriented categories such as resource guilds, wetland types, or microhabitats.

3.2.1 DIVERS

Divers (or **loons** as they are known in North America) of the order Gaviiformes are large, northern forest (Common Loon) or Arctic tundra species (four other species) well adapted to feed on fish, amphibians, and large invertebrates by long, spear-like bills and webbed feet placed far back on the body. They often feed in deep water, can dive to depths of more than 75m, and can change body density by compression of body and feathers so they can sink slowly. As a result of diving adaptations, they walk poorly on land and rarely do so. Feeding and nesting tend to be solitary, although Red-throated Loons are social. Nest sites are selected on fairly solid substrates immediately adjacent to water, sometimes enhanced with vegetation from the nest area along shores of islands in freshwater lakes. Red-throated Loons often build nests on small wetlands or pools on islands and fly to adjacent larger bodies of water for food, which is carried in their bill to the young at the nest area. Clutches of eggs tend to be small, and the young are precocial and down-covered and able to swim shortly after their down is dry. Loons are day-time migrants and move long distances from northern breeding areas to warmer wintering areas on large freshwater lakes, marine bays, or coastal waters. After breeding and probably after fall migration, a complete wing molt occurs on lakes selected for sufficient size and food resources to ensure survival during the flightless period.

3.2.2 GREBES

Grebes (order Podicipediformes), like the loons, are skilled diving specialists but are a more diverse group of about 21 species worldwide in distribution. However, they are most abundant in cold-temperate climates occurring through either latitude or altitude. All have lobed toes for swimming underwater with great maneuverability, and their terminally placed legs make it necessary for them to crawl on their bellies on land. Like loons, they can sink underwater without active jump-diving, as is common in ducks and coots. Most species of grebe favor more shallow and often well-vegetated wetlands rather than great expanses of open and deep water used by loons. Their bills vary from short and sharp to long and spear-like, dependent upon their aquatic prey. They nest mainly in freshwater wetlands, sometimes in sizable colonies, and may build nests of submergent vegetation in open water or of decomposing to fairly fresh emergent vegetation in quite dense marsh vegetation. The precocial young tend to spend a few days on or near the nest, where they are fed by the parents; some are fed until nearly grown and feeding independently. Some species may be double-brooded (i.e., attempt to rear two clutches per breeding season) where temperature and water conditions permit. Adults and young eat fish, amphibians, and large invertebrates. They migrate long distances at night to winter in large freshwater lakes or marine bays. Three high-altitude Central/South America species (or subspecies) endemic to large lakes are essentially flightless.

3.2.3 PELICANS

The Order Pelecaniformes contains a diverse group of large-bodied waterbirds most commonly associated with nearshore marine systems. However, many are freshwater species and a few are found almost entirely there. Their feet are uniquely adapted for swimming, with webbing between all four toes (toti-palmate), rather than between three (palmate), which is common to divers and ducks.

The seven or eight species of **pelicans** occur on all major continents and are the best known of the order Pelecaniformes because of their large bill and "gular" pouch. Most species capture fish, amphibians, or other aquatic prey in water while surface swimming or tipping up. The relatively smaller Brown Pelican is the only one that regularly fishes by aerial plunge-diving with closed wings from 20m or more above the water, like the related oceanic **boobies** (six species) and **gannets** (three species). Brown Pelicans frequent tropical to warm temperate nearshore marine waters whereas the other species nest in inland fresh or saline lakes in the Americas, Africa, and Australia, but they may winter along seacoasts in highly saline waters. Most nest socially on the ground on islands, although Brown Pelicans may nest in shrubs or trees such as mangroves. The young are altricial and naked and are fed in or near the nest until feathered and able to swim. Adults may travel up to 75km to gather food (Findholt and Anderson 1995).

3.2.4 SHAGS AND CORMORANTS

Shags (about 14 species) and **cormorants** (23 species) are worldwide members of the Pelecaniformes; the former subfamily is especially well represented in marine and fresh cold waters at high latitudes whereas the cormorants are more tropical and freshwater (Johnsgard 1993). One tropical cormorant of the Galapagos Islands is flightless. All species feed mainly on fish, are skilled aquatic divers, and often feed in mixed flocks with pelicans and gulls. Most species are social nesters, often mixed with herons, egrets, ibises, and anhingas in colonies in tree, shrub, or marshy nest sites. Coastal birds nest on cliffs socially or in mixed flocks of seabirds. The young are altricial, naked at hatching, and remain in the nest when small. Most are strongly migratory, often moving in goose-like wedges even in local moves from overnight roosts to feeding areas.

3.2.5 ANHINGA AND/OR DARTERS

The four species of anhinga and/or darters are tropical or subtropical birds of dominantly freshwater wetlands (Portnoy 1977); they are shag-like in shape but with longer necks, straight bills, and long, fan-shaped tails. They are some-times called "snakebirds" because they are skilled divers and can sink in water like grebes, with only the head and neck exposed. Solitary feeders, they take mostly fish by impaling with one or both mandibles rather than by grabbing (Johnsgard 1993); amphibians and large aquatic invertebrates also are eaten. They frequent shallow and often vegetated wetlands but seem to favor shrub or open-forest areas where they can readily perch. They have difficulty in getting airborne and may climb a perch first, but they then fly strongly with a dis-tinctive "flap-and-glide" technique. Like pelicans and frigatebirds, they readily soar. Nests typically are over water in shrubs, small trees, or snags and are found with those of other colonial waterbirds such as herons, ibises, and storks. The altricial young are fed by both parents on regurgitated food. After nesting, adults move to larger water bodies (Meanley 1954), presumably where they undergo a complete and simultaneous wing molt (the only pelecaniform to become flightless during molt)(Palmer 1962). Birds that nest in warm-tem-perate areas seem to migrate to tropical wintering areas (Palmer 1962).

3.2.6 FRIGATE BIRDS

The five species of frigatebirds are mainly tropical oceanic seabirds that often nest on remote islands but also may nest or roost in shrubs and small man-grove trees along coastlines and, therefore may be considered birds of coastal wetlands. They are the ultimate aerialists, gracefully surface-skimming for flying fish or snatching squid, rarely if ever landing on water and may even drink from freshwater pools on-the-wing. They commonly "pirate" food from boobies, gulls, and other fish or squid feeders. They nest colonially in bushes or low mangroves, commonly in areas attracting other species of nesting seabirds.

3.2.7 WATERFOWL: DUCKS, GEESE, AND SWANS

The waterfowl (wildfowl in Europe) or ducks, geese, and swans of the order Anseriformes are a large and cosmopolitan group widely distributed in trop-ical as well and extreme north and south latitudes (Delacour 1954–64, Johnsgard 1981). They are more conspicuous than many abundant birds because of size, their day as well as night migration, and their large and conspicuous flocks. Species that live at high latitudes are strongly migratory and are, therefore, excellent pioneers at exploring and exploiting new habitats. Several endemic species of subantarctic and cold-temperate islands are flightless (Weller 1980).

Nest sites vary from cliffs and tree holes to highly aquatic but well-vegetated situations. The young are precocial, downy, and leave the nest to feed with other brood members as soon as they are dry. Young birds are strongly insectivorous except in the true herbivores like geese and swans. Parental care (brooding and predator defense) ranges from long term (geese and swans) to none (the parasitic Black-headed Duck). After breeding, males in particular but sometimes females as well engage in molt-migration, flying a few to hundreds of miles to suitable feeding areas with protective water depths where they lose their flight feathers prior to autumn migration (Hochbaum 1944). A few species molt after post-breeding migration.

Waterfowl, and especially the ducks, are best grouped by larger taxa for discussion because such groupings tend to reflect common names of the groups and describe their behavior and resource use. The term tribe, a taxonomic category between subfamily and genus, will be used here (Table 3.1) because it has been popular and functional and is still a basis for the latest classification although subfamily status has been given to some tribes (Livezey (1986).

Screamers of South America (three species) are large, goose-like birds with chicken-like bills and semipalmated feet; they graze on meadows, wet grasslands, and shallow marshes. The Southern Screamer nests over water in robust emergent vegetation, and the precocial young feed on aquatic invertebrates and plants until able to feed safely in drier areas. They are taxonomically distinct from all other members of the order and are usually placed in a separate suborder (Anhimae) and family (Anhimidae).

Although a unique-looking bird, the single species of **Magpie** or **Pied Goose** is more closely related to the suborder of typical waterfowl but is placed in a separate family (Anseranatidae). It occurs in tropical Australia where it feeds on plant tubers and other vegetation by wading or upending; it is highly social in feeding as well as nesting and is unique among the suborder Anseres in having a serial wing molt. Frith (1967) considered this an adaptation to the drying of the marshy swamps in which it nests.

Whistling ducks number about nine species; they are closely related to swans and are of tropical to warm-temperate distribution around the world in diverse habitats. Some field-feed in farm crops and nest in tree holes in semi-desert (Black-bellied) and others dive for invertebrates in densely vegetated subtropical pools (Fulvous); this results in seasonal and regional variation in food and in foraging tactics. The White-backed Duck of Africa often is placed with or near the whistling ducks as a separate tribe or subtribe, but it is a highly aquatic diving specialist. The seven or so swan species (sometimes separated as the tribe Cygnini) are well known as usually white aquatic birds that feed on submergent vegetation as first choice but will also graze in meadows. The Black

Plate I. Wetland types and vegetation structure.

(*a*) The Prairie Pothole Region of southern Manitoba is made up of small to medium sized herbaceous Palustrine wetlands in an agricultural setting. Despite loss of wetland acreage and natural water regimes through drainage, such areas produce diverse waterbirds because of the mix of size, depth, and plant succession.

(*b*) Palustrine forested and shrub bottomland wetlands seasonally flooded by early summer rains. Ideal habitat for breeding, Prothonotary Warbler and White-eyed Vireo (Bethel, Texas).

(*c*) A large beaver-impounded wetland showing the typical shallow edge zonation of lotus, rushes and sedges, shrubs, and adjacent water-tolerant trees. Used by North American Wood Duck, Common Moorhen, and various herons and egrets as feeding and nesting sites during summer (Bethel, Texas).

(*d*) Intensive taro cultivation on Kauai Island in Hawaii has resulted in excellent feeding and roosting habitat for the Koloa or Hawaiian Duck, which has suffered population declines from drainage and development elsewhere on the island.

(*e*) Brown Teal searching for invertebrates or resting on an estuarine gravel or shingle shoreline. At high tide, teal move into riverine pools or upland areas to feed (Great Barrier Island, New Zealand).

(*f*) Red Mangroves in an estuaine impoundment that forms cover for crabs and fish, which are food resources for Great Egret, Wood Stork, and White Ibis (Sanibel, Florida).

(*g*) Meandering mountain stream demonstrating riffles, meanders, and pools influenced by beaver dams. Habitat for Mallards, Black-crowned Night-Herons and shrub warblers and flycatchers (Independence Pass, Colorado).

(*h*) Riverine habitat with gravel bars and adjacent meadows in Tierra del Fuego used by Ashy-headed Sheldgeese and Chiloe Wigeon as brood-rearing and feeding habitat.

(*i*) The Kissimmee River of central Florida was straightened and deepened for flood control, with great loss of meander channels and backwater habitats. Managed natural habitats show the goals of restoration efforts now in progress.

(*j*) In a setting more suitable for nesting penguins and other seabirds, these peat-based freshwater ponds produce enormous numbers of fairy shrimp, which are eaten by resident South Georgia Pintails that have pioneered and reared their young under these sea-moderated conditions. Such wetlands presumably are enriched by the nutrients that result from upslope nesting of myriads of seabirds (South Georgia).

(*k*) Intermountain basins often collect water and form saline sinks that are productive of invertebrates and attract flamingos and ducks (Abra Pampa, Argentina).

**Plate II. Feeding tactics
and sites.**

(*a*) Auckland Island Flightless
Teal feeding on armored
isopods by rooting in stranded
and decomposing kelp (Adams
Island).

(*b*) Upending or tipping
Whooper Swans, a common
feeding method used also by
many species of dabbling ducks
to reach the wetland substrate
when feeding in deeper water
(Welney, UK).

(*c*) Laysan Duck or Teal using the "run-snatch" tactic for capturing shore flies swarming along the shoreline of the main hypersaline lagoon on Laysan Island. This technique also is used by some shorebirds.

(*d*) Crested Screamer feeding in water velvet (General Lavalle, Argentina).

(*e*) Kelp Sheldgoose feeding at low tide on sea lettuce from partially exposed rock strata (Falkland Islands).

(*f*) Tricolored Heron in the strike position after waiting in a camouflaged site in cattail, followed by a quick but cautious walk to prey location (Aransas Pass, Texas).

(g) Great and Snowy egrets and a mixed-flock of shorebirds feeding in their preferred water depths in a drying pool (Corpus Christi, Texas).

(h) Common or European Coot searching for and surface-pecking for floating food items (London Zoo).

(*i*) Mixed flock of feeding Neotropic Cormorants and American White Pelicans (Port Aransas, Texas).

(*j*) Atlantic coast sand beach where Red Knots, Ruddy Turnstones, and Laughing Gulls feed on spawn of Horseshoe Crabs (Cape May, New Jersey).

(*k*) Least Bittern with fish captured adjacent to a walkway of a sewage lagoon nature preserve (Port Aransas, Texas)

(*l*) Black-necked Stilts in open estuarine sheetwater (Corpus Christi, Texas).

(*m*) Although the South Georgia Pipit nests in the upland grass like other pipits, it spends much of its foraging time in the decomposing seaweed along the shore, which is rich in crustaeans (Albatross Island, South Georgia).

(*n*) Cattle Egrets and other waterbirds that feed on insects and other prey use African Elephants and many other large herbivores as "beaters" (Amboseli National Park, Kenya).

(a) Water-soaked Anhinga sunning and "wing-spreading" in Red Mangroves (Sanibel, Florida).

(b) Puna Ibis preening while sunning, showing its long, curved bill which is well designed for probing in soft mud (London Zoo).

(*c*) Although flightless, wing-flapping on water by this Auckland Island Flightless Teal is a common comfort movement, as it is in most birds, it helps to shed water and rearrange feathers (Adams Island).

Swan of Australia, and the Black-necked Swan and Coscoroba Swan (with black wing tips) of temperate South America, contrast with the all-white swans of the northern hemisphere. Marine and estuarine vegetation is used by swans in both hemispheres.

Geese (tribe Anserini) number about 16 species throughout the northern hemisphere and are specialized as grazers in meadows and uplands but may feed in wet areas on tuberous plants. The more coastal Brant feed on eelgrass or algae in estuaries during migration and in winter and may use grasses and sedges during breeding on freshwater lakes. Some wild species have been introduced widely for hunting, and domesticated varieties of Eurasian geese are important human food resources worldwide.

A complex tribe (or subfamily (Livezey (1986)) includes **shelducks** and **sheldgeese** (commonly called geese because of body shape and feeding behavior but not taxonomically the same as northern hemisphere geese). These show a great range of habitat selection and specialization: alpine meadows (Andean Goose), clinging marine algae (Kelp Goose), and tropical savannah (Egyptian Goose and South American Orinoco Goose). The shelducks are more duck-like in body form but also may graze in moist grasslands or feed on estuarine or freshwater invertebrates as well (Common Shelduck).

The four species of steamer-ducks found at the tip of South America often are classified with shelducks but are massive diving specialists forming an ecological equivalent of the northern hemisphere eiders. The three flightless species are found only on marine coasts, whereas the flying species frequents large inland lakes.

Dabbling ducks (Anatini) number between 45 and 51 species and are considered the most ubiquitous and adaptable of all waterfowl and among the most widespread of all waterbirds. Species complexes such as mallard- and pintail-like ducks span both hemispheres and are successful in a great variety of habitats from cold latitudinal and alpine tundra to the tropics. They are flexible in foods, pioneers in new habitats of even temporary water, and are persistent in rearing young under tenuous conditions. Foods of these omnivores include seeds and other plant reproductive parts during nonbreeding seasons and invertebrates during pre-breeding, growth, and molting periods. Nests commonly are on the ground near water, but clutches are large and re-nesting is common.

Bay ducks or pochards (Aythyini) include 17 species and are nearly as widespread as dabblers but require more permanent waters, where they feed seasonally on invertebrates and submergent plant material. Breeding areas usually are on freshwater, but wintering birds may use either saline or freshwater. Some members of this group nest over water in emergent vegetation, but ground nests near water are common in other species.

Sea ducks (Mergini) such as eiders and goldeneyes number about 20 species and are found mainly in the northern hemisphere. They are well designed for cold water regardless of salinity; most eiders nest near tundra ponds and some winter and feed among Arctic ice floes. Other sea ducks like goldeneyes and mergansers nest in tree holes in coniferous forests and winter in large lakes and river systems, often in warmer climates. Harlequin Ducks, goldeneyes, and mergansers are well adapted to fast-moving streams where they feed on invertebrates (Harlequin) or fish (goldeneyes and mergansers). Long-tailed Ducks rival loons in their diving ability.

The Stifftail ducks (Oxyurini) include nine species. They are skilled divers of worldwide distribution and frequent shallow but often open water (Johnsgard and Carbonell 1996). Their long and often upright tail is used in social displays but undoubtedly is also a useful diving tool. They feed mainly on benthic invertebrates for much of the year, although submergent foliage and seeds are used during winter in some areas. The young are among the most precocious of all birds; the young of the parasitic Black-headed Duck leave the host's nest a day or two after hatching and are thereafter independent (Weller 1968).

3.2.8 FLAMINGOS

Flamingos (order Phoenicopteriformes) are widely known because of their typically vivid pink coloration, unique bill shape, feeding behavior, unusual body form yet graceful flight, and their highly social behavior. The Lesser Flamingo specializes on algae (*Spirulina* spp.)(Zimmerman, Turner and Pearson 1996), the Greater Flamingo seems to favor microcrustaceans and mollusks, but the American Flamingo also uses some bulbs and seeds of emergent plants (Arengo and Baldassare 1995). These are obtained from bottom oozes by straining through their sieve-like bill with the head upside-down in the water or mud; the bill is flattened laterally rather than dorso-ventrally as in spoonbills. Foot-stirring is common to enhance food intake, as is true in a slightly different form in fish-feeding egrets. Flamingos build unique cone-like nests of mud – often quite high to avoid flooding – on mudflats in highly saline and sometimes caustic lakes (pH of up to 10.5). African populations have demonstrated highly nomadic patterns in search of suitable breeding areas after traditional areas have been flooded or totally dried (Crowther and Finlay 1994). Others remain in certain areas and often face nest failure or juvenile mortality because of drying habitat and lack of freshwater. They rear few young per nesting effort but seem to live long lives (40 or more years). The young are precocial and down-covered but tend to remain in the nest for a week or more and are fed with a regurgitant until fledged. Their bill shape changes rapidly with growth, and they are able to strain water for food in an adult-like manner in about a month. Except for island populations, flocks may engage in both migration to wintering areas and more local molt-migration prior to becoming flightless (as in ducks and geese and a few other families or orders).

The order **Ciconiiformes** (Table 3.1) is a large, diverse, and taxonomically complex group of herons, egrets, and related groups. They have worldwide distribution and are among the more visible and attractive wetland birds. Most are characterized by long legs suitable for wading (termed waders in the USA whereas this term is used for shorebirds in Eurasia). Prominent, long bills are characteristic of the group and differ in length and shape through feeding adaptations. Many of these species and groups feed in mixed flocks opportunistically when fish, crustaceans, amphibians, or other prey become vulnerable owing to declining water and oxygen levels. Some members of this order are solitary or loosely social nesters (bitterns and storks, Green Herons) but most are colonial, often in mixed company with anhingas or cormorants. Although many species nest over water or at least near water at elevated sites, many nest on land on islands, and a few like bitterns may nest in upland vegetation when ground predators are absent (Duebbert and Lokemoen 1977). Nest materials are carried some distance or robbed from old nests or other birds' nests. Because parents can bring food long distances, they can nest far from water on artificial structures (Great Blue Herons on power-line structures) or groves of trees (many species of herons and egrets). Young are altricial and only sparsely covered with fluffy down; they are fed by regurgitation.

Herons (including **egrets** and **bitterns**) have long, straight bills designed for grasping and holding fish, amphibians, reptiles, and sometimes birds and mammals. Bill sizes and shapes vary in herons and egrets, presumably representing adaptations to their usual prey size and behavior. Moreover, prey size of adults generally is correlated with body size. Some are normally solitary feeders (Great Blue Heron, Grey Heron, bitterns) and others regularly feed in social groups when capturing prey in food-rich pools (Great, Snowy, and Cattle Egrets).

Spoonbills are the invertebrate-gleaners of the order, able to eat small crustaceans, insects, and mollusks as well as fish and amphibians by virtue of a shoveller-like spatulate bill. However, they can and do eat large fish opportunistically. The highly social **ibises**, some of which are highly terrestrial, have long but down-curved bills suitable for probing in mud and sweeping or groping in water to contact and grab prey like fish or large invertebrates. They often are attracted to muddy flats or sheetwater resulting from flooding and therefore, commonly feed and move as flocks even locally. Some differences in food use has been noted with some species taking rice and other seeds (Acosta *et al.* 1996).

Storks are the largest members of the order and have broader and sometimes seemingly disproportionate bills effectively designed for large and especially uncooperative prey, including rodents, snakes, and amphibians as well as fish. Some are mainly terrestrial for much of the year; others like Wood Storks feed mostly on aquatic vertebrates in water. White Storks build huge nests in tree snags and even chimneys as nest sites, and White Storks and Marabou Storks supplement their food from refuse dumps (Blanco 1996, Zimmerman *et al.* 1996). The unique African Hamerkop feeds in both water and on land but builds a stick nest, which it enters through an opening in the side.

3.2.10 HAWKS AND EAGLES

Hawks and eagles (order Falconiformes) are mostly facultative or opportunistic wetland users, but several that specialize on aquatic foods are almost restricted to wetlands: Snail Kites of North and South America feed almost exclusively on Apple Snails and nest over water in small shrubs. Ospreys, nearly worldwide in coastal areas and large rivers and lakes, feed mainly on fish in shallow waters but build nests in snags and artificial structures near water. The African Fish Eagle, Bald Eagle of North America, and several sea-eagles of Eurasia are adept at taking fish and not afraid of getting wet; although they are somewhat flexible in habitat, they seem to favor seacoasts, large lakes, or rivers, where they often feed on dead, stunned, or vulnerable live fish. Like Ospreys, eagles often build large nests, which may be used year-after-year; the young of all members of the order are altricial. A few hawks favor open marshy wetland edges and nest on the ground in fairly wet places (harriers) but also hunt in and occasionally nest in upland grasses and forbs. Some forest hawks favor wet shrub and bottomland wetlands (South American Black-collared Hawk, North American Swallow-tailed Kites and Red-shouldered Hawks) or mangroves but are not restricted to these habitats (Howell and Chapman 1997). Similar responses to water are noted among other species of kites worldwide.

3.2.11 RAILS, CRAKES, COOTS, AND CRANES

The rails and crakes, coots, and cranes of the large, diverse, and worldwide order Gruiformes range from the more conspicuous cranes and coots to the habitat-shielded rails of emergent vegetation. Many rails and crakes frequent the shoreward zone in dense vegetation, where they are more commonly heard than observed. Moorhens, gallinules, and coots use deeper and even open water. Foods are highly variable, with seasonal shifts from animal foods (especially invertebrates) to seeds, tubers, and foliage in fall and winter. Coots regularly use open water when feeding on submergent vegetation, and they dive for both vegetation and invertebrates – especially during pre-nesting egg formation and when feeding young. They are equipped with lobate toes effective for diving as well as for walking on mudflats. Coots and probably other members of this family lose the power of flight during their complete and simultaneous wing molt. Rails fly little except when pressed but are capable long-distance migrants and inhabit some of the most remote islands of the world. Several island species are flightless or nearly so, as is the famous Takahe or Notornis of

New Zealand, which occupies tussock swamps and steep but soggy slopes above 1000 m elevation with little standing water.

Cranes switch from nesting areas in marshes, where they eat amphibians and other small vertebrates, to open grasslands (tubers) or fields in winter (grains). Some cranes (Whooping) use estuarine areas and feed on crabs and other marine invertebrates but also take freshwater crayfish and acorns opportunistically. Most species are found in the northern hemisphere, where they nest in bogs and marshes after long daytime migrations in goose-like flocks, vocalizing as they travel. Sandhill Cranes rest during migration on river islands and sheetwater, where night-time roosts are adjacent to fields in which high-energy foods are readily available. The more terrestrial species seem to have prospered as a result of current grain crops, but southern forms have lost nesting and roosting sites to intensive agriculture and water developments.

The **Limpkin** of Florida and Central and South America has some ibis-like features such as a decurved bill and general stature but is a close relative of the cranes and rails. It is larger than ibises, is typically well-hidden in marsh vegetation, and its screaming call alerts one to its unseen presence. It feeds mainly on Apple Snails and other large invertebrates. Nests are in marshy areas or low shrubs, and the young are precocial and downy, leaving the nest soon after hatching as in other members of the order. Although commonly solitary, it may occur in flocks.

Although the three Finfoots seem to be related to the rails, they have grebe-like bills and feet. One species is often called the **Sungrebe.** Both the species found in tropical America and Africa favor slow-moving streams (Alvarez del Toro 1971, Zimmerman *et al.* 1996), swim well, and are more fish-oriented than others in this order; they also eat crustaceans and large insects. Most nest sites are elevated over water or at the water's edge; the young are nearly precocial and are fed in the nest before ranging afield. Adults apparently walk well on land and wade as well as dive in shallow water. They take flight with difficulty, favor well-vegetated areas, may roost in trees at night, and are not very conspicuous. The single species of **Sunbittern** of Central and South America is a stream-edge wader, nesting in solitary pairs along streams, and feeding on aquatic resources. The young are precocial, as are others of this order.

3.2.12 CHARADRIIFORMES

The order Charadriiformes is a large, diverse, complex, and important group involving about 14 mostly obligate taxa (families, subfamilies or tribes) of wetland birds best identified by common group names. Best known are the gulls and terns, sandpipers and associates like bog-country snipe, forest-dwelling woodcock, phalaropes, stilts and avocets, and plovers. The auks and allies also are members of this order but are truly seabirds and nest colonially on rocky islands and cliffs. One member of the group, the Marbled Murrelet, nests on the horizontal branches of old-growth conifers growing near the northwest coast of North America and eastern Asia. Most members of the order are strong fliers and interhemisphere long-distance migrants; the spectacular migration and complex routes of Red Knots and Golden and Grey Plovers between Arctic nesting areas and southern continents are renowned (Harrington 1996).

Plovers, especially the three Golden Plovers, share with sandpipers the strategy of long-distance migrations; which allows them to tap global resources. They are more terrestrial than the sandpipers, feeding and nesting in drier and shorter vegetation but commonly use wetland shorelines. The common North American Killdeer and the threatened Piping Plover are well known: the former tolerating human development to the point of nesting on gravel rooftops; the latter frequenting temporary habitats often eliminated by development and reservoirs designed for water management. The young are precocial as is common to species of dynamic habitats. The Glareolidae of Eurasia includes the **coursers** and swallow-plovers or **pratincoles**. They are plover-like and nest on the ground. The Collared Pratincole nests on dry meadows or cropland (Calvo and Furness 1995) and hawks for aquatic insects over wet meadows and marshes (Calvo 1996). The coursers are more terrestrial, but one member of that group, the Egyptian Plover, nests and feeds along sandbars of large rivers. The Crab-plover, the only member of the family Dromadidae, is strictly a coastal species; its name describes its food (Fasiola, Canora and Biddar 1996) but it is unique in its petrel-like behavior of nesting in burrows in sand dunes, often in huge colonies (Campbell and Lack 1985).

The sandpiper family is large and diverse and often separated into tribes. **Sandpipers** are birds of shallow, open or grassy sheetwater, mudflat, or sandy shore; they feed on live invertebrates, spawn, or small fish or amphibians most of the year. Many are habitat-specific, with preferences for substrate, water depth, wave height, wind, etc., so that adaptations of leg length, neck length, bill length, and shape differ by species and sometimes by sex. They glean, peck, pick, drill, bore, probe, and sweep, and they demonstrate efficient adaptations in behavior as well as in structure. Their nesting habitats often differ markedly from their winter or migration feeding areas, and one sandpiper, the Solitary Sandpiper, nests in old songbird nests in stunted trees in the northern forest edge that is interspersed with grassy feeding pools (Kaufman 1996). The young are precocial. **Curlews** are larger species with long, down-curved bills and longer legs; some feed on mudflats and other species favor uplands. They may nest in wet-meadows or tall grasses and winter in fresh or saline coastal wetlands where at times they feed like Willets or avocets while wading. **Woodcock** specialize in moist forested or shrubby bog wetlands, where they probe for earthworms and other burrowing invertebrates of the northern hemisphere.

The **oystercatchers** number six or so species of truly coastal shorebirds that are recognized by their black or black and white plumage and straight, sturdy, and brightly colored chisel-like bills. They are worldwide in distribution except for high latitudes; they frequent rocky to sandy shores and feed on limpets, clams, and other shellfish.

The **avocets** and **stilts** are considered either as a separate family or as a tribe of shorebirds. They are worldwide in distribution and are characterized by long legs, useful in shallow open water. They favor saline/soda wetlands and lake shores or estuaries. Stilts have long straight bills and seem more commonly to be running on mudflats than wading, but that tactic depends on feeding opportunities. Avocets have recurved (i.e., upturned) bills ideally suited to sweeping the substrate while wading in shallow water up to belly-deep; in fact they sometimes swim. They often nest on sites with little vegetation, and the precocial young are well camouflaged for their habitat.

The **phalaropes** are among the most aquatic of shorebirds; they may behave like other walking or wading shorebirds but more commonly they swim like feathered "whirligigs," feeding on tiny crustaceans by bill dipping. The food seems to be held on the bill by water tension, and the mouth is hardly opened to take in the food (Rubega and Obst 1993). They nest by or in fresh-water wetlands in temperate or Arctic areas, using relatively short emergent vegetation like water-tolerant grasses or sedges, near and sometimes over water and, therefore, have food nearby. This group is characterized by reverse sexual plumage colors and sexual roles, with males doing most of the incubation and care of young. They are strongly migratory, with the Red (or Grey in Europe) and Red-necked Phalaropes often wintering at sea. All three species regularly migrate in autumn to winter in tropical and southern hemisphere habitats and are very tolerant of saline waters.

Jacanas are the tropical or subtropical "water-walkers;" they have light frames, long toes to disperse the body weight, and can gracefully flit across even submergent vegetation and especially over lily pads and similar floating-leaf vegetation. They feed mainly on snails, insects, and other invertebrates. Rather small and shallow nests are made of plant material on floating or rooted vegetation, and eggs are incubated by one of several males associated with the same laying female. The eggs are incubated by the males, and the precocial young leave the nest shortly after drying. Young may be chased or killed by new females entering the mating group (Emlen, Demong and Emlen 1989).

Many of the common gull-like birds are grouped as one family but make up at least four subfamilies. Gulls and terns, while closely associated and often nesting together or on similar sites, represent two subfamilies that reflect different feeding strategies involving adaptations of flight and feeding apparatus. Terns are predators on live fish or aquatic invertebrates taken on or near the surface of fresh- or saltwater, and a few capture insects in flight over land. Most are skilled at hovering, plunge-diving, and devouring their prey aloft. Black Terns skim marsh water during breeding or dry fields in migration for insects.

Gulls are more commonly scavengers or kleptoparasites ("hijackers"), harassing and taking food from other birds. During the breeding season, marsh-nesting species like Franklin's Gulls in North America and Brown-hooded Gulls in Argentina feed dominantly on marsh flies (Chironomidae). Marsh-nesting terns like the Whiskered Tern may nest on water-lily pads, whereas the Black Tern typically nests on soggy and rotting plant debris but may build a nest of fresh plants on a floating structure. Others, especially along large lakes or seashores, nest on islands or shorelines on sand or rocks with little nest material. The Forster's Tern does both, varying with region and habitats (marsh, lakeshore, or seashore). The young of gulls and terns typically are precocial but stay in or near the nest while being fed insects or bits of regurgitated fish, crabs, and squid.

Jaegers and **skuas** are gull-like but are placed in a separate family or sub-family; they are typically coastal marine birds but they often nest on land near freshwater areas. They are large and more predators than scavengers, taking mammals, fish, or young birds or eggs during the breeding season.

Skimmers are placed in a separate family because of their unique adaptations for surface feeding, including an extended lower bill and elegantly designed wings for surface flight. They feed on fish and large invertebrates by skimming with the lower bill and grabbing on contact, by day or night. Although considered tactile feeders, they seemingly are adapted for nocturnal vison by a larger than usual number of cones in the retina (Rojas, McNeil and Cabana 1997). They commonly nest on bare to mat-covered sandbars along the coast (Burger and Gochfield 1990) or on large lakes and river systems, where they can easily take flight. As a result, where locally abundant, they are commonly seen at nests, and some populations even use gravel rooftops for nesting.

3.2.13 CUCKOOS

There are few cuckoos that frequent wetlands but several parasitic species do lay eggs in over water nests of other birds. The Old World Swamp Cuckoo or Pheasant Coucal and the Mangrove Cuckoo in subtropical America (sometimes considered a race of the Yellow-billed Cuckoo) were named for the common habitat selection. But one unique species, the South American Hoatzin, was once taxonomically linked with gallinaceous birds but now is considered an aberrant cuckoo by most taxonomists. It nests over water in small trees along slow-moving streams and feeds on coarse vegetation of wetland trees, absorbing their nutrients through a foregut fermentation system (Gralal 1995). Young remain in the nests until able to clamber around in the branches, aided by vestigial claws on the wings.

3.2.14 OWLS

Although few people in America and Europe think of Owls as wetland birds, a few are well adapted for preying on animal food resources of rivers and basin wetlands. There are three species of fishing-owls of Africa and four fish-owls of Asia, which specialize on fish much as Ospreys and eagles do, but they often feed at dusk or night. They lack extensive feathering on the legs and have gripping feet to snatch fish from the surface. They also eat frogs, crabs, and other large invertebrates. The widespread and mostly high-latitude Short-eared Owl frequents marshy meadows of grass and sedge as well as grasslands and nests on the ground, sometimes in habitats as wet as that of harriers. Its food is mainly rodents and small birds, and its association with water sometimes makes it a predator of other wetland birds (Holt 1994). Although the widespread Short-eared Owl does not nest in Africa, a close relative, the African Marsh Owl, nests in wet vegetation and seems to take even more wetland prey. A similar habitat is used by a member of the Barn Owl family, the African Grass Owl, which nests on the ground in wet or damp places. The Barred Owl of the southern USA favors mature and wet bottomland woods along large rivers, where it feeds on amphibians, fish, and crayfish in addition to mammals and birds.

3.2.15 KINGFISHERS

Kingfishers are a worldwide group of about 87 species (Fry, Fry and Harris 1996) that are far more diverse in habitat choice and feeding habits than generally appreciated. There are three families, of which only two are perch-and-dive fish eaters. The largest family of about 56 species comprises mainly forest hunters like the Australian Laughing Kookaburra and the Shovel-billed Kingfisher or Kookaburra of New Guinea, which gropes on the forest floor in search of grubs and worms. The traditional wetland kingfishers (31 species in two families) specialize on fish or large invertebrates of various sizes, with prey body size often related to bird body size (Remsen 1990). They use a variety of wetland types, including saltwater, where perches allow. Many are stream-oriented, and artificial ditches or canals are favored sites, especially when wires are conveniently situated. Others frequent pools of all sizes, and although they seem to favor perches of certain heights (Remsen 1990), they can hover effectively some distance from shore; for example, the Pied Kingfisher of Africa has been recorded some 3km from shore on large lakes (Fry *et al.* 1996). Like many fish-feeders, they capitalize on prey trapped during declining water levels so will use roadside ditches and tiny pools far from other water. Most nest in burrows or crevices, rear few young per brood, and vary greatly in migration patterns.

3.2.16 SONGBIRDS

Compared to most of the taxa mentioned above, few of the perching birds or songbirds (order Passeriformes) are obligate wetland specialists with anatomical, behavioral, and physiological adaptations, but many tap water resources and nest at the water's edge. Most probably evolved from adjacent upland species that successfully reared young from nests built in shoreward wetland vegetation because of the protection from predators afforded by the water, but also because of the richness of invertebrate foods essential to growing nestlings and common to wetlands. Most have altricial young in cup, dome, hole, or crevice nests, and feeding by parents is essential until young are flying.

Several passerines are adapted behaviorally to streams or rivers. In the mountainous parts of the northern hemisphere and in South America, there are five dippers that walk, swim, dive, and feed in, and nest along, flowing mountainous streams. They seem tied to several insect groups like caddisflies that frequent these areas. A number of South America ovenbirds favor marshy areas, and the genus *Cinclodes* uses, streamsides or seashores. The African Slender-billed Starling is associated with waterfalls, hopping around and feeding in the spray, and nesting behind the falls even in groups (Zimmerman *et al.* 1996).

Several wrens of North America are well adapted to fresh and saline marshes, with the Marsh Wren found in no other habitats and the Sedge Wren using shallow marshes of less-robust, grassy vegetation or wet prairie. This habitat-use pattern applies also to several New World grackles and smaller blackbirds, mentioned above. Additionally, each area of the world seems to have a few marsh, rush, reed, or shoreline specialists among flycatchers, vireos, finches, and warblers. The last range in North America and include Prothonotary Warblers nesting in tree-hole cavities in bottomland swamps, Palm Warblers of northern bogs, Bachmann's (Remsen 1986), Swainson's, and Hooded Warblers of southeastern cane breaks, and Yellow Warblers, which seem always near water but use diverse sizes of shrubs and trees for nesting. The New World warblers called waterthrushes and the ovenbirds use small habitat patches along stream courses or lowland woods. A number of warblers use mangroves either for feeding or wintering, but little is known about their relative use of these versus adjacent habitats. The closely related Common Yellowthroats also frequent water areas and may nest in emergent plants like cattail but most often use upland shrubs and thickets. As noted elsewhere, there seem to be many Eurasian and African warblers (family Sylviidae) that specialize in emergent marshes where no Icterids or wrens compete.

Flycatchers are fairly common residents of shrubs and emergents at the water's edge; the Alder and Willow Flycatchers of North America and the Swamp Flycatcher in Africa are essentially restricted to such habitats. Although the Kiskadees of subtropical areas use a wide range of habitats, they generally frequent water areas where they feed on dragonflies or large invertebrates and fish.

Few seed-eating sparrows have obvious morphological specialization for wetlands, but many use them and a few seem limited by habitat patterns and feeding behavior (see Rising 1996). Some, like Song Sparrows, seem most abundant in floodplains, where they use small trees for song perches and brush for nesting, but they have wide latitude in habitat use. Several sparrows use damp to wet vegetation of grass and sedge meadows (LeConte's and Henslow's Sparrows). Lincoln's Sparrow favors northern boggy habitats. Only a few actually nest in wet, marsh-edge emergent herbaceous plants or small trees (Swamp Sparrow), but more species feed in wet or shallowly flooded emergent vegetation (Nelson's Sharp-tailed Sparrow) even far from shore (Seaside Sparrows). Populations of several species of both more upland (Savannah) and wetland-edge species (Song and Swamp Sparrows) have races or sibling species that nest in tidal saltmarsh (e.g., Powell 1993).

Swallows feed over and may nest over water but despite a strong association with wetland insects lack specializations that restrict them to wetlands. Tree Swallows and Mangrove Swallows feed mainly over water and wetland vegetation, presumably because of the abundance of food (St Louis, Breebart and Barlow 1990), and tend to nest in holes in snags over or near water.

3.2.17 SUMMARY

The use of wetlands and their resources is widespread among many diverse bird taxa of the world, with some using them to the exclusion of other habitats via specializations that maximize exploitation of the rich resources but that limit use of alternative habitats. Many groups adapted to other habitats have few wetland representatives occurring only regionally or locally and are limited in adaptation to behavioral responses such as selection – perhaps as a product of reproductive success.

Hopefully, this sketchy coverage will introduce readers to the elaborate adaptations and gradients of wetland use demonstrated by many taxonomically diverse birds of the world and encourage additions to this list.

References

Acosta, M., Mugia, L., Mancina, C., & Ruiz, X. (1996). Resource partitioning between Glossy and White Ibises in a rice field system in Southcentral Cuba. *Colonial Waterbirds* 19, 65–72.

Alvarez del Toro, M. (1971). On the biology of the American Finfoot in Southern Mexico. *Living Bird* 10, 79–88.

Arengo, F. & Baldassarre, G. A. (1995). Effects of food density on the behavior and distribution of nonbreeding American Flamingos in Yucatan, Mexico. *Condor* 97, 325–34.

Blanco, G. (1996). Population dynamics and communal roosting of White Storks foraging at a Spanish refuse dump. *Colonial Waterbirds* 19 273–9.

Burger, J. & Gochfield, M.(1990). *The black skimmer: Social dynamics of a colonial species.* New York, NY: Columbia University Press.

Calvo, B. (1996). Feeding habitats of breeding Collared Pratincoles (*Glareda pratincola*) in Southern Spain. *Colonial Waterbirds* 19 (Spec. Publ. No. 1): 75–7.

Calvo, B. & Furness, R. W.(1995). Colony and nest-site selection by Collared Pratincoles (*Glareda pratincola*) in southwest Spain. *Colonial Waterbirds* 18, 1–10.

Campbell, B. & Lack, E. (1985). *A dictionary of birds.* London: T & AD Poyser.

Crowther, G. & Finlay, H. (1994). *East Africa.* Hawthorn, Victoria: Lonely Planet Publishing.

Delacour, J. (1954–64). *The waterfowl of the world*, Vols. 1–4. London: Country Life.

Duebbert, H. F. & Lokemoen, J. T. (1977). Upland nesting of American Bitterns, Marsh Hawks, and Short-eared Owls. *Prairie Naturalist* 9, 33–40.

Emlen, S. T., Demong, N. J., & Emlen, D. J. (1989). Experimental induction of infanticide in female Wattled Jacanas. *Auk* 106, 1–7.

Fasiola, M., Canora, L., and Biddar, L. (1996). Foraging habits of crab plovers *Dromas ardeola* overwintering on the Kenya coast. *Colonial Waterbirds* 19, 207–13.

Findholt, S. L. and Anderson, S. H. (1995). Foraging areas and feeding habitat selection of American White Pelicans (*Pelecanus erythrorhychos*) nesting at Pathfinder Reservoir, Wyoming. *Colonial Waterbirds* 18, 47–68.

Frith, H. J. (1967). *Waterfowl in Australia.* Honolulu, HI: East- West Press.

Fry, C. H., Fry, K., and Harris, A. (1996). *Kingfishers, bee-eaters and rollers.* Princeton, NJ: Princeton University Press.

Gill, F. B. (1995). *Ornithology*, 2nd edn. New York: W. H. Freeman.

Gralal, A. (1995). Structure and function of the digestive tract of the hoatzin (*Opisthocomus hoatzin*), a folivorous bird with foregut fermentation. *Auk* 112, 20–8.

Harrington, B. (1996). *The flight of the red knot.* New York: W. W. Norton.

Hochbaum, H. A. (1944). *The Canvasback on a Prairie Marsh.* Washington, DC: Wildlife Management Institute.

Holt, D. W. (1994). Effects of Short-eared Owls on Common Tern colony desertion, reproduction, and mortality. *Colonial Waterbirds* 17, 1–8.

Howell, D. L. & Chapman, B. R. (1997). Home range and habitat use of Red-shouldered Hawks in Georgia. *Wilson Bulletin* 109, 131– 44.

Johnsgard, P. A. (1981). *Ducks, geese, and swans of the world.* Lincoln, NB: University of Nebraska Press.

Johnsgard, P. A. (1993). *Cormorants, darters, and pelicans of the world.* Washington, DC: Smithsonian Institution Press.

Johnsgard, P. A. & Carbonell, M. (1996). *Ruddy ducks and other stifftails.* Norman, OK: University of Oklahoma Press.

Kaufman, K. (1996). *Lives of North American birds.* New York, NY: Houghton Mifflin.

Livezey, B. C. (1986). A phylogenetic analysis of recent Anseriform genera using morphological characters. *Auk* 103, 737–54.

Meanley, B. (1954). Nesting of the water turkey in Eastern Arkansas. *The Wilson Bulletin* 66, 81–8.

Palmer, R. S. (1962). *Handbook of birds of North America,* Vol. 1. New Haven, CT: Yale University Press,

Portnoy, J. W. (1977). *Nesting colonies of seabirds and wading birds – coastal Louisiana, Mississippi, and Alabama.* OBS/FWS77–07. Washington, DC: US Fish & Wildlife Service,

Powell, A. (1993). Nesting habitat of Belding's Savannah Sparrows in coastal salt marshes. *Wetlands* 13, 219–23.

Remsen, J. V. Jr (1986). Was Bachman's Warbler a bamboo specialist? *Auk* 103, 216–19.

Remsen, J. V. Jr (1990). Community ecology of neotropical kingfishers. *University of California Publications in Zoology* 124, 1–116.

Rising, J. D. (1996). *A guide to the identification and natural history of the sparrows of the United States and Canada.* San Diego, CA: Academic Press.

Rojas, L. M., McNeil, R., and Cabana, T. (1997). Diurnal and nocturnal visual function in the tactile foraging waterbirds: the American White Ibis and the Black Skimmer. *Condor* 99, 191–200.

Rubega, M. A. and Obst, B. S. (1993). Surface-tension feeding in phalaropes: discovery of a novel feeding mechanism. *Auk* 110, 169–78.

St Louis, V. L., Breebaart, L., and Barlow, J. C. (1990). Foraging behavior of Tree Swallows over acidified and non-acid lakes. *Canadian Journal of Zoology* 68, 2385–92.

Weller, M. W. (1968). The breeding biology of the parasitic Black-headed Duck. *Living Bird* 7, 169–207.

Weller, M. W. (1980). *The island waterfowl.* Ames, IA: Iowa State University Press.

Zimmerman, D. A., Turner, D. A., and Pearson, D. J. (1996). *Birds of Kenya and northern Tanzania.* Princeton, NJ: Princeton University Press.

Further reading

Ali, S. (1979). *The book of Indian birds.* Bombay: Bombay Natural History Society.

Ali, S. & Ripley, S. D. (1983). *A pictorial guide to the birds of the Indian Subcontinent.* Delhi: Oxford University Press.

Allen, A. A. (1914). The red-winged blackbird, a study in the ecology of a cat-tail marsh. *Proceedings of the Linnaean Society of New York* 24, 43–128.

Allen, J. N. (1980). The ecology of the Long-billed Curlew in southeastern Washington. *Wildlife Monographs* 73, 1–67.

Austin, O. (1961). *Birds of the world.* New York: Golden Press.

Bellrose, F. C. (1980). *Ducks, geese and swans of North America,* 3rd edn. Washington, DC: Wildlife Management Institute.

Bent, A. C. (1919–68). *Life histories of North American birds.* Washington, DC: US National Museum.

Bildstein, K. (1993). *White Ibis: wetland wanderer.* Washington, DC: Smithsonian Press.

Brown, L. H. and Urban, E. K. (1969). The breeding biology of the Great White Pelican *Pelecanus onoerotalus roseus* at Lake Shala, Ethiopia. *Ibis* 111, 199–237.

Brown, L. H., Urban, E. K., and Newman, K. (1982). *The birds of Africa,* Vol. I. London: Academic Press.

Burger, J. (1985). Habitat selection in temperate marsh-nesting birds. In *Habitat selection in birds,* ed. M.L. Cody, pp. 253–81. Orlando, FL: Academic Press.

Blanco, G. (1996). Population dynamics and communal roosting of White Storks foraging at a Spanish refuse dump. *Colonial Waterbirds* 19, 273–9.

Burton, J. A. (1971). *Owls of the world.* New York: E. P. Dutton.

Clancey, P. A. (1967). *Game birds of southern Africa.* New York: American Elsevier.

Curson, J., Quinn, D., and Beadle, D. (1994). *Warblers of the Americas.* New York: Houghton-Mifflin.

Cuthbert, N. L. (1954). A nesting study of the Black Tern in Michigan. *Auk* 71, 36–63.

Daiber, F. C. (1982). *Animals of the tidal marsh.* New York: Van Nostrand Reinholt.

DeSchaunsee, R. M. (1971). *A guide to the birds of South America.* Edinburg: Oliver and Boyd.

Eddleman, W. R., Evans, K. E., and Elder, W. H. (1980). Habitat characteristics and management of Swainson's warbler in Southern Illinois. *Wildlife Society Bulletin* 8, 228–33.

Erlich, P. R., Dobkin, D. S., and Wheye, D. (1988). *The birder's handbook.* New York: Simon and Schuster.

Erskine, A. J. (1971). *Buffleheads.* Monograph Series No. 4. Ottawa: Canadian Wildlife Service.

Evans, P. R., Goss-Custard, J. D., and Hale, W. G. (eds.) (1984). *Coastal waders and wildfowl in winter.* Cambridge: Cambridge University Press.

Faaborg, J. (1976). Habitat selection and territorial behavior of the small grebes of North Dakota. *Wilson Bulletin* 88, 390–9.

Falla, R. A., Sibson, R. B., and Turbott, E. G. (1967). *A field guide to the birds of New Zealand.* Boston, MA: Houghton Mifflin.

Ferns, P. N. (1993). *Bird life of coasts and estuaries.* Cambridge: Cambridge University Press.

Gibb, J. (1956). Food, feeding habits and territory of the Rock Pipit. *Ibis* 98, 506–30.

Glover, F. A. (1953). Nesting ecology of the Pied-billed Grebe in northwestern Iowa. *Wilson Bulletin* 65, 32–9.

Gonzales, J. A. (1996). Breeding biology of the jaribu in the llanos of Venezuela. *Wilson Bulletin* 108, 524–34.

Gorena, R. L. (1997). Notes on the feeding habits and prey of adult Great Kiskadees. *Bulletin of the Texas Ornithological Society* 30, 18–19.

Hamerstrom, F. (1986). *Harrier, hawk of the marshes.* Washington, DC: Smithsonian Press.

Hancock, J. A. (1992). *Storks, ibises, and spoonbills of the world.* San Diego, CA: Academic Press.

Hancock, J. and Kushlan, J. (1988). *The herons handbook.* New York: Harper Row.

Harrison, C. (1978). *A field guide to the nests, eggs, and nestlings of North American birds.* London: Collins.

Hori, J. (1964). The breeding biology of the Shelduck *Tadorna tadorna. Ibis* 106, 333–60.

Johnsgard, P. A. (1981). *Plovers, sandpipers, and snipes of the world.* Lincoln, NB: University of Nebraska Press.

Johnsgard, P. A. (1983). *Cranes of the world.* Bloomington, IN: Indiana University Press.

Johnsgard, P. A. (1987). *Diving birds of North America.* Lincoln, NE: University of Nebraska Press.

Johnsgard, P. A. (1990). *Hawks, eagles, and falcons of North America.* Washington, DC: Smithsonian Institution Press.

Kear, J. & Duplaix-Hall, N. (1975). *Flamingos.* Berkhamsted, UK: T & AD Poyser.

Kilgo, J. C., Sargent, R. A., Chapman, B. R., and Miller, K. V. (1996). Nest-site selection by Hooded Warblers in bottomland hardwoods of South Carolina. *Wilson Bulletin* 108, 53–60.

Kirsh, E. M. (1996). Habitat selection and productivity of Least Terns on the lower Platte River, Nebraska. *Wildlife Monographs* 132, 1–48.

Madge, S. & Burn, H. (1988). *Waterfowl, an identification guide to the ducks, geese and swans of the world.* Boston, MA: Houghton Mifflin.

Martin, A. C., Zimm, H. S., and Nelson, A. L. (1951). *American wildlife and plants: a guide to wildlife food habits.* New York: McGraw-Hill.

McCabe, R. A. (1991). *The little green bird: ecology of the Willow Flycatcher.* Madison, WI: Rusty Rock Press.

Monroe, B. L. Jr and Sibley, C. G. (1993). *A world checklist of birds.* New Haven, CT: Yale University Press.

National Audubon Society (1994). *Pocket guide: familiar birds of lakes and rivers.* New York, NY: Alfred A. Knopf.

National Audubon Society (1997). The ninety-fifth Christmas bird count. *National Audubon Society Field Notes* 51, 135–710.

Nettleship, D. N. & Duffy, D. C. (eds.) (1995). The Double-crested Cormorant: biology, conservation and management. *Colonial Waterbirds* 18 (Special Publication No. 1), 1–256.

Niering, W. A. (1985). *Audubon Society Nature Study Guides: Wetlands.* New York, NY: Alfred A. Knopf.

Olsen, K. and Larsson, H. (1994). Terns of Europe and North America. Princeton, NJ: Princeton University Press.

Orians, G. H. (1980). *Some adaptations of marsh-nesting blackbirds.* Princeton, NJ: Princeton University Press.

Orians, G. H. and Horn, H. S. (1969). Overlap in foods and foraging of four species of blackbirds in the Potholes of central Washington. *Ecology* 50, 930–8.

Ortega, B., Hamilton, R.B., and Noble, R.E. (1976). Bird usage by habitat types in a large freshwater lake. *Proceedings of the Southeastern Fish & Game Conference* 13, 627–33.

Owen, M. (1977). *Wildfowl of Europe.* London: Macmillan.

Paulson, D. (1993). *Shorebirds of the Pacific Northwest.* Seattle, WA: University of Washington Press,

Pizzey, G. (1980). *A field guide to the birds of Australia.* Princeton, NJ: Princeton University Press.

Provost, M. W. (1947). Nesting birds in the marshes of northwest Iowa. *American Midland Naturalist* 38, 485–503.

Reed, C. A. (1965). *North American birds eggs;* revised edition. New York, NY: Dover.

Reicholf, J. (1975). Biogeographie und Okologie der wasservogel im subtropisch-tropischen Sudamerika (with English summary). *Anzeiger der Ornithologischen Gesellschaft in Bayern* 14:1–69.

Ripley, S. D. (1977). *Rails of the world.* Boston, MA: David R. Godine.

Ryan, M. R., Renken, R. B., and Dinsmore, J. J. (1984). Marbled godwit habitat selection in the northern prairie region. *Journal of Wildlife Management* 48, 1206–18.

Scott, D. A. & Carbonell, M. (1986). An annotated checklist of the waterfowl of the Neotropical realm. In *A directory of neotropical wetlands*, compilers D. A. Scott and M. Carbonell, pp. 669–84. Cambridge and Slimbridge, UK: International

Union for Conservation of Nature and Natural Resources and International
Waterfowl Research Bureau.

Sealy, S. G. (1978). Clutch size and nest placement of the Pied- billed Grebe in
Manitoba. *Wilson Bulletin* **90**, 301–2.

Sibley, C. and Monroe, B. (1990). *Distribution and taxonomy of birds of the world*.
New Haven, CT: Yale University Press.

Sibley, C. and Monroe, B. (1993). *A supplement to distribution and taxonomy of birds
of the world*. New Haven, CT: Yale University Press.

Tacha, T. and Braun, C. (1994). *Migratory shore and upland game bird management
in North America*. Washington DC: International Fish and Wildlife Agencies.

Tuck, L. M. (1972). *The snipes: a study of the genus* Capella. Monograph Series No. 5.
Ottawa: Canadian Wildlife Service.

Tyler, S. and Ormerod, S. (1994). *The Dippers*. San Diego, CA: Academic Press.

van Tyne, J. & Berger, A. J. (1976). *Fundamentals of ornithology*. New York, NY: John
Wiley.

Weller, M. W. (1961). Breeding biology of the least bittern. *Wilson Bulletin* **73**, 11–35.

Weller, M. W. (1972). Ecological studies of Falkland Islands' waterfowl. *Wildfowl* **23**,
25–44.

Weller, M. W. and Fredrickson, L. H. (1974). Avian ecology of a managed glacial
marsh. *Living Bird* **12**, 269–91.

Weller, M. W. and Spatcher, C. E. (1965). *Role of habitat in the distribution and abun-
dance of marsh birds*. Iowa State University Agriculture and Home Economics
Experiment Station Special Report No. 43, pp. 1–31. Ames IA: Iowa State
University Press.

4

Water and other resource influences

Habitat needs for all birds include similar resources: water for drinking and bathing, food, cover for protection from the elements and potential predators, and undisturbed space for meeting social and other life functions. But each group of birds adapts to general features of their typical habitat and exploits its particular resources. Obviously, the unique and dominant feature for wetland birds is water, and long-term adaptations include genetically selected modifications in anatomy, morphology, and physiology (Table 1.1). Shorter-term adjustments involve mostly behavioral changes, but we know little about these and they are difficult to study and test experimentally. Birds, like many wetland animals, often are not truly water adapted but retain terrestrial adaptations that allow them to survive with extremes of water regimes and also to exploit resources of both habitats along the wetland edge.

Migratory, or at least mobile, birds obviously have greater flexibility in their search for habitat than do flightless, terrestrial vertebrates. It has been suggested that birds logically should follow a hierarchy of selection from large to smaller scale. First, on the largest scale, they either reside in or migrate to large geographic regions with climatic regimes and general landscape or vegetation features attractive to them (either via instinctive preferences or prior experience). Second, a bird could select from various landscape alternatives within this geographic region, such as areas showing certain types, spatial relationships, or sizes of wetlands. Third, a single wetland or complex that provided suitable habitat might be the focus of the bird's search for that season or function (e.g., breeding, nonbreeding, feeding, roosting), and, finally, selection of microhabitats, foraging sites, nest site, etc. within the chosen wetland or wetland complex will be made to meet its needs. Both the general region and the specific area presumably would meet some innate habitat image, possibly influenced by experience, but some trial-and-error exploration may be involved – especially in cases of resource shortages.

General needs will be considered first here and more detailed examples will be given throughout the text. I will start with water because wetland birds are influenced by this constantly, and in ways that few of us may be aware.

4.1 Water uses and influences

Water as a resource can be for drinking, bathing, and escape from predators. We accept the facts that birds need water and that its availability is less of a problem for waterbirds than for more terrestrial and especially desert birds. But it is the psychology of water's presence that seems to be of first importance in water-adapted birds. It is an attractant, and no habitat is habitat for these species without visually significant bodies of water. Use of a habitat as a source of suitable foods is the next priority, followed by use for body care, such as bathing, thermoregulation, etc. Bathing is regular and often is a group activity. Sometimes it is incorporated into courtship behavior in obligate waterbirds that normally mate in and nest on, over, or near water.

Many wetland birds can swim, but some do so rarely. There is a scale of use of water among wetland birds. Facultative species like Common Yellowthoats (Kelly and Wood 1996) and Swamp Sparrows use wetland edges opportunistically and may feed and nest over water rarely. Most shorebirds are always associated with water but some barely get their feet wet on mudflats; the hasty scampering of Sanderling running before the waves is an amusement to many but it is geared more to feeding tactics than getting wet – going where the clams are! True waterbirds that swim regularly vary from birds of quiet waters like Moorhens or dabbling ducks to open-water bay ducks like Redheads and Scaup, which often seek the shelter of inland ponds or protected bays when wind and wave velocities exceed a certain level. Such ponds may also provide fresh water (Adair, Moore and Kiel 1996). It is simply less work to avoid the waves unless they are feeding, in which case they feed and leave. Skilled swimmers and divers like grebes and loons can hardly walk on land because of their terminally placed legs, but they may be found in large open bays and deep depths where they are unseen from shore. Only the sea ducks like Long-tailed Ducks, scoters, and eiders swim and dive in big expanses of deep and cold water. Down obvious plays an important role in both insulation and keeping dry and is most dense under the contour feathers of the breast, the area most exposed to water. Even in calmer wetlands, many birds seem to avoid both breezes and waves. Is this because they do not like the discomfort of ruffled feathers, or is it because of thermoregulation? It seems obvious that compact plumage should reduce energy loss. Birds such as terns and frigatebirds rarely sit on water, perhaps a good strategy to avoid fish or seal predation, but it probably has more to do with the need to stay airborne to find food. Moreover, their plumage does not seem well designed for extending swimming either, and loafing is on an exposed bar, rock, or tree.

Selection of depth of water for feeding is an important and poorly studied behavior, mainly because it is not easy to measure quickly at a distance and perhaps because its significance in habitat segregation has not been appreciated. But each species seems to have a pattern of use geared to the foods and the area, and most are fairly consistent. Some of these tactics vary with season and

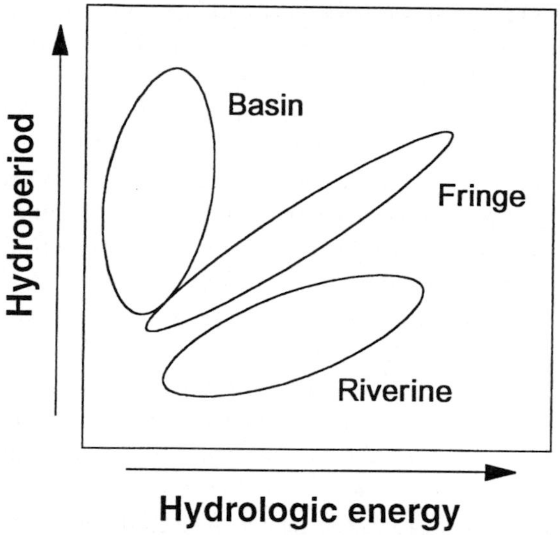

Figure 4.1. Relationship between typical water regularity (here termed hydroperiod) and hydrologic energy in the three major wetland types in the hydrogeomorphic classification method (Brinson 1993).

food. For example, avocets are among the more aquatic shorebirds, usually wading in water deep enough that they can sweep their upturned bills from side to side in search of food. But when the water levels are deep as a result of flooding or tides, they swim and probe head-under like a duck. When that occurs, it is not unusual for other shorebirds like Greater Yellowlegs and Willets to be feeding in deeper-than-normal water too, presumably induced by food resources.

Although water attracts waterbirds and provides both food and protection, it has its dangers because it can be a powerful force. It may directly harm or kill, or stress birds through energy demand, thus influencing the species adaptations and ultimately the species composition of birds using certain wetland types. Based on general observations, it seems that the hydrologic forces of the three major types of wetland (Fig. 4.1, Brinson 1993) typically differ in their bird species richness. During the breeding season in North America, it is common to find 15 to 25 obligate species in productive basin wetlands, perhaps 8 to 12 species in fringe wetlands, but many fewer species in river channels. This pattern presumably results both from energy demands of feeding in such systems and from the nature and abundance of suitable resources present, which in turn is the product of the physical stresses that influence the basic productivity of each system: current in streams and wave action in larger lakes and oceans. These stresses affect water regimes, which in turn influence plant physiology (Riemer 1984, Schulthorpe 1967) and plant succession processes (which will be elaborated below).

Except for periodic inflow from sheetwater or stream overflow, basin wetlands lack current and have the least wave energy except as produced by wind. Obviously, the larger the wetland the greater the wind fetch and potential for wave energy to build. Although such wetlands have vertical water movement resulting in water-level changes (Brinson 1993), birds are less influenced by current or wave action. Therefore, these basins are used by a wide variety of birds for breeding and feeding, with less potential for impact of waves on nests, young, or food resources. Many of these species are efficient at building up nests during flooding, and others built nests high in the vegetation, which minimizes loss. Where protective vegetation is lost in larger wetlands between years, wave action increases the potential of loss of nests and eggs and hampers feeding efficiency, with a net loss of energy on the part of the bird. As a result, the habitat becomes less optimal for breeding but might serve other purposes, such as feeding or roosting during calm periods or for species that have different feeding tactics (e.g., underwater diving by grebes as opposed to surface feeding by dabbling ducks). When water levels decline, new dangers of predation may develop, and abandonment of nests and wetlands is common, having the potential result of saving the adults rather than losing both adult and nest.

Potentially serious wave action reaches far greater magnitude along the fringes of large lakes, coastal bays, and oceans, where both wind and tidal energies are expressed as enormous waves that have life-threatening potential and must be avoided by even well-adapted birds. Therefore, shorebirds like sandpipers, oystercatchers, and plovers avoid the waves, and birds like cormorants, grebes, and loons feed offshore rather than in the wave wash. Some bay ducks (e.g., Greater Scaup) and most sea ducks like eiders and scoters cautiously use large wave-washed shores, but some mortality has been recorded at such sites. The large northern eiders (Bustnes and Lonne 1995) and the massive flightless steamer-ducks of Patagonia probably are the best adapted birds for this zone, but they often use the cover of kelp leaves where waves are quieted rather than facing an unnecessary energy demand, although even their downy young can do so (Weller 1980).

Streams and rivers demonstrate tremendous variability in volume and rate of flow and thereby area of available wetland or deeper aquatic habitat. In smaller, arid-land streams, this can be an erratic and spectacular event that changes the amount of habitat dramatically (Stanley, Fisher and Grimm 1997). Large river systems reflect larger watersheds and, therefore, flooding potential (Miller and Nudds 1996) and can have tremendous impact on riverine wetlands by influencing water levels, by destroying some basins via sedimentation, and by creating other wetlands by scouring action. For birds that use marginal or riparian vegetation for nesting and feeding, direct impact is perhaps less common except at extreme flood stages. Those that are more aquatic in both feeding and breeding may be immediately more attracted to the higher water stages but may face potential impacts of extreme current and unpredictable water levels.

River systems demonstrate linear variation in stream width, volume, and velocity, which varies from high to low gradient. As mentioned above, stream ecologists classify these variations according to current speed and channel

Water and other resource influences

width, e.g., small and intermittent, narrow and high velocity, torrent and riffle, pool and riffle, rapid and broad, meandering and slow, etc. Obviously, all rivers act like torrents during flood events, but torrents and other high-energy riverine systems regularly would be the most energy demanding locality for birds; therefore, we would expect few bird species (or fish or invertebrates for that matter) because of the adaptations required. It is not surprising that most of the birds that use streams feed on caddisflies and similar stream insects, but they also eat fish eggs. Once such adaptations are acquired, stream specialists are essentially limited to such habitats, which is less true of birds of basin wetlands. We will discuss in Chapter 15 the implications of human-induced changes in stream speed and configuration as it might affect dippers in alpine areas (Tyler and Ormerod 1994) or ducks and herons in slower stream systems (Weller 1995).

Adaptations of many birds to feeding in swift water include a laterally compressed body, powerful legs placed well back on the body, and often long bills suitable for probing between gravel and rocks, and perhaps auditory adaptations to water noise, which otherwise could induce higher predation. Many of these birds reduce the impacts of current by diving and feeding underwater, and the current against the back and long tails of Torrent Ducks, Common Mergansers, and Harlequin Ducks provide downward and forward thrust (comparable to putting your flat hand out of the car window to experience this free energy in an air medium). Obviously, tail length for use in water must be optimized with use in flight. The only diving passerines of high-energy streams, the dippers, do not have long tails, but some authors point out that the entire back functions as a driver when the head is down. In fact, dippers seem to have very few adaptations for what they do so well! Dippers seemingly use their wings to swim underwater and then walk or swim according to need. They also avoid the high-velocity portion of streams by feeding along the shoreline, and several South American species dive little. I have also seen Andean Torrent Ducks swim up-current along the edges of a flooded stream, where current was less but where no human could stand, but even Torrent Ducks may fly short distances to avoid this effort.

There are other stream specialists among ducks that offer an interesting comparison: the Blue Duck of New Zealand (Eldridge 1986, Kear 1972) probably is the nearest to the Torrent Duck in current tolerance, but it tends to avoid streams with great volume of high-speed flow. The Spectacled Duck of the lower Andes chooses slower, broader, and more shallow streams and may also use small basin lakes, as does the African Black Duck (McKinney *et al.* 1978, Siegfried 1968). The North American Wood Duck is considered a riverine bird, but it favors backwater areas where it nests in large trees in bottomland woods; most feed and rear their young in quiet, backwater areas or even ponds (Bellrose and Holm 1994). Finally, Canada Geese may graze along large, slow-moving streams or streams with pools, but they tend to nest on islands or adjacent cliffs away from floods and predators.

Ice is a form of water that often is used as a resting area briefly or overnight and for escape cover, but it creates problems for divers in pools and lakes, inhibits feeding by dabblers when the bottoms are still frozen (which can last for weeks or months after surface water melts in northern wetlands), and limits access to food. Starvation has been documented among coots during early spring migration when weather changes reduced food access and stressed the fat reserves of the birds (Fredrickson 1969). Recently, satellite-tracking of radio-tagged eiders has shown that they use pack ice areas in the north much as penguins do in the southern hemisphere, loafing on ice and feeding by diving in the icy waters between the floes (Petersen, Douglas and Mulcahy 1995). Undoubtedly, we have much more to learn about winter life at sea for the many species of ducks and phalaropes that use freshwater wetlands during the breeding season.

Tidal action in coastal wetlands is a product of celestial gravitational influences but also of meterological conditions such as wind and barometric pressure. These influence water depth and food availability both in saltwater areas and in estuarine and freshwater tidal marshes, often in unpredictable and unexpected ways. Tidal action in saltwater areas influences availability of foods in fringing wetlands such as rock substrates (Weller 1980), seagrass meadows (Cornelius 1977), or mudflats (Withers and Chapman 1993). As a result, birds either do not feed during high-water periods or move into alternative habitats such as ponds or truly upland feeding areas for New Zealand Brown Ducks (Weller 1974), or flooded agricultural fields for shorebirds (Rottenborn 1996). Few of us think about the upslope influence of tides upon freshwater flowing into the sea, but a discrete plant community is formed there, and birds in that area are fairly diverse but relatively low in population density compared with saltwater areas (Odum *et al.* 1984). Water levels in these wetlands are influenced by the coastal tidal regime, the distance to the sea, and the volume of freshwater inflow. Such levels have a major impact on nesting waterbirds that have water-level nests, such as rails and ducks, but also influence passerines that nest higher in tall emergent grasses like *Spartina* spp. For example, Seaside Sparrows not only suffer high levels of nest loss during high spring tides but also have high re-nesting potential, which results in the synchrony of nesting populations with lower tidal regimes later in the season, and, in ultimately, higher nest success (Marshall and Reinert 1990).

4.2 Salinity

Although most birds seem to have salt-removal glands as well as suitable excretory processes, birds vary in their ability to tolerate high salt intake, ranging from those that perhaps can but rarely do to those oceanic birds that rarely drink freshwater or have salt-free foods (albatrosses, petrels, and auks).

Moreover, there is considerable evidence that increased use improves the functioning of salt glands under high salinity (Bradley and Holmes 1972, Cornelius 1977).

North America White and Brown Pelicans often segregate by use of nesting islands in inland freshwater or saline lakes versus coastal saltwater. Degree of salinity may be a factor but prey availability is more important. Predator protection also must be important, however, because the colony of White Pelicans in the Great Salt Lake in Utah nests where there are few large prey items available and makes long flights to food sources to feed their young. White Pelicans of African soda lakes must get along with no or limited freshwater, but they and young flamingos do seek out and use freshwater for drinking and bathing, which may well be essential for their survival. While many flamingos use only high-saline waters, they may favor saltier water for some functions and avoid it for others; patterns of use demonstrate that factors such as food, water depths, and substrate influence this choice (Espino-Barros and Baldassarre 1989). Salinity gradients seem easily recognized by birds as Laysan Ducks often go to the seashore to bathe rather than using a hypersaline lagoon (Moulton and Weller 1984).

Among those groups of birds that commonly nest in or near freshwater during the breeding season, several eiders use only freshwater or estuary nest sites, but Common Eiders will use full-strength seawater (about 35 parts per thousand (p.p.t.)). Although adults ducks regularly drink saltwater, they seem more able to tolerate this heavy salt load than can juveniles, and therefore, breeding often is on freshwater rather than salt – to a point that it may well limit the distribution of many species (Schmidt-Nielsen and Kim 1964, Weller 1980). In alkaline lakes in arid regions, efficient salt glands are important survival mechanisms even though we think of the birds using these lakes as freshwater birds (Cooch 1964). Many waterfowl species in marine environments nest at stream outlets where they can get freshwater rather than in other seemingly suitable situations. Wintering birds feeding in hypersaline areas often make regular flights to freshwater ponds to drink (Adair *et al.* 1996, Woodin 1994). Moreover, studies of body condition in Northern Shovelers showed that birds wintering on saltwater sites were less fat than those of most freshwater sites (Tietje and Teer 1988).

Recent studies of White Ibis (Bildstein 1993 and others cited therein) have demonstrated that birds nesting in coastal areas may feed many miles inland to obtain foods that are less salt-laden than those found along the coast. If young are fed excessive salt loads, growth is reduced and their condition is generally less healthy (Figure 4.2). It is clear that salt costs the bird energy in physiological removal of salt or in flight energy to seek freshwater. In cases where the physiology of the species is not well adapted to high salt intake, the alternative is selection of food with a reduced salt content. Obviously, the latter may increase adult vulnerability to predators and other dangers arising from mobility.

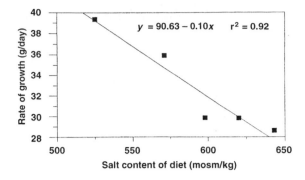

Figure 4.2. Growth rate of young White Ibis in relation to the salinity of water in which their food was taken (Bildstein 1993, with permission).

4.3 Food

Food is so important to birds on a continuing basis that we will devote the next chapter to this topic, but for perspective, it is important here to identify patterns by group and to clarify that food needs differ greatly by stage of the life cycle, making generalizations difficult. Except for those birds specialized as herbivores (specialized gut structure and complex fermentation mechanisms), most omnivores need higher animal protein during egg development, growth of the young, and body molt (Eldridge and Krapu 1988, Heitmeyer 1987; Krapu 1974) They need high-energy carbohydrates during the physical stress of migration and during thermoregulation at low winter temperatures. As a result, their diets differ seasonally and by function, and birds move, seek, and vary food choices to meet these needs. Food for young often differs from that of adults unless the adults are in molt or body-growth phases. Young omnivores gradually shift from animal protein in early growth to more seeds and then foliage as they mature. Carnivores or piscivores show shifts more in size or species of prey, often dictated by selection made by the parents during the feeding phase.

Getting food of the correct type and at the right time is one issue, but some birds are not effective until they have the proper feeding conditions. Hence, a feeding perch is preferred or needed by most kingfishers or flycatchers. Aerialists like swallows or plunge divers like Ospreys or Brown Pelicans have greater flexibility, but even these require areas large enough to permit flight maneuvers, and water of sufficient depth to reduce potential injury. For most species, there are habitat-related issues such as water depth, water clarity, and vulnerability of the prey. Lack of disturbance by potential predators (including people) also may be important.

4.4 Resting

All birds need undisturbed places for rest both during the day and at night to avoid predation or stress. If one observes at a site where various waterbirds fly

Water and other resource influences

in during the day, some land in water and start to drink, some start to feed, and some land on a mudflat or snag, often disturbing other individuals or species to find a place to rest. Some species have very specific requirements for rest sites, presumably based on genetic makeup and social structure. Many swimmers like ducks simply float on open water, but losses could occur through predation by large fish or alligators, and, therefore, birds are always "testing the waters." Many rest along bare shorelines or mudflats, preferably on islands where they are protected from ground predators at night, and the benefits of social tolerance produce alertness for predators. Vulnerability probably is greater at nights, but vocalizations of geese, cranes, and ducks suggest that sleep is not as sound as in humans. Some species like coots apparently sit on mats of submergent vegetation where alligators cannot reach them effectively. Cranes (Tacha, Nesbitt and Vohs 1994) and many geese and Southern Screamers in South America need open sheetwater to roost overnight. For wintering Lesser Snow Geese on the Gulf of Mexico coastal plain, use of a region seems to depend upon the availability of such a roost area, and birds may come and go at night if feeding conditions or disturbance patterns demand.

Many opportunistic wetland users such as swallows or blackbirds perch on emergent vegetation or trees over water, but they are subject to predation by owls, mink, and perhaps snakes. Ground roosters like Northern Harriers may be subject to predation by Great-horned Owls (Weller, Adams and Rose 1955), but still larger perchers like herons and Ospreys use snags or posts in conspicuous places but are large enough to escape aerial predators.

4.5 Thermoregulation

Because of their high body temperature, thermoregulation may seem of less importance to birds than to other animals, but there is strong evidence that, in addition to migrating to warmer regions during cold periods, they select microhabitats to reduce chilling winds, overtopping waves, or other stressful situations (Burger *et al.* 1984). Observers may chill at the sight of ducks splashing around in a pool in an ice-edged wetland, forgetting that water temperatures are above freezing within the water and are, therefore, a comparatively warmer site in what may be a very cold ambient temperature. Waterfowl are especially able to winter in cold areas because of their compact feathering and heavy down layer. Canada Geese can even manage for weeks or months without water if they have high-energy food resources like grains in the northern USA. Energetic costs of various types of activity are better understood now (Wooley and Owen 1978) and are related to how birds maintain body temperature. Mallard ducks may remain buried in snow or other heat-preserving shelters rather than go out to feed on severely cold days – presumably saving more energy than they could take in (Jorde *et al.* 1984). Black Ducks wintering in coastal Maine conserve heat behaviorally by tucking their feet into flank feathers and their bill into scapular feathers (Brodsky and Weatherhead 1984), and by selecting sheltering landforms (Longcore and Gibbs 1988).

4.6 Escape cover

The need for and use of predator escape cover or conditions vary greatly among groups and species. Most of us immediately think of the swimming waterbirds that can and do dive when pursued from above or within the water, but waders and walkers gain surface protection only by the presence of water, and the ability to fly from ground predators. Anyone who has observed waterbirds for long has seen bobcats, coyotes, or foxes attempt such predation, but their success at catching ducks seems low except during nesting – hence the high mortality of females, especially common in ground-nesting species. It is particularly important that rearing areas have water and food to sustain the young during their flightless period, which can be quite long in geese and swans.

4.7 Habitat requirements for breeding

Birds also need some very specific habitat features for reproduction, and this varies greatly with the taxonomy and habitat choice of the bird group in question. Breeding for some taxa like songbirds and herons requires territorial sites to proclaim ownership and attract mates (Lack 1968). There are important genetic and evolutionary implications to such habitat choices by males; male Red-winged Blackbirds (a polygynous species) that sing in low-quality habitats may not attract as many females as those that select better nesting areas with food for the young. In colonial waders, territory may be the preliminary nest site where the male has deposited enough sticks or other vegetation to create a mating platform, but the site may be some distance from food. Ducks need open-water areas for on-the-water courtship displays that establish and test pair bonds, much of which occurs on wintering and migration areas. Terns engaging in fish flights need not only a fish to pass to their prospective mate, but undisturbed beach areas where ceremonial feeding occurs. It seems obvious that the prevalence of food-oriented breeding displays accomplishes not only pairing bonds but tests the habitat for resources, but there is little direct proof of this.

Despite the intensive study of the breeding biology of many of the waterbirds, and the obvious importance of understanding nest sites as ingredients of reproductive success, they are among the more difficult features to study. We have produced lots of descriptions, measured parameters, assessed range of locations, and yet we have difficulty predicting whether a wetland will attract birds to build nests there (see Knopf and Sedgwick 1992). Among wetland birds, nest sites typically are found in microhabitats ranging from highly aquatic to terrestrial: (i) completely floating nests of buoyant vegetation (small grebes); (ii) in water but essentially resting on some substrate (some rails and ducks); (iii) above water and remote from shore (Least Bitterns, herons); (iv) near shore but at wet-to-damp sites (some rails, American Bitterns, and ducks); (v) underground burrows of mammals or natural crevices (some goldeneyes population); (vi) dry ground with varying degrees of short, herbaceous cover, at various distances from, but associated with, water (Common Yellowthroats, Sedge Wrens, some ducks); (vii) at bases of tall emergent vegetation, over land or water (sparrows, some New World blackbirds); (viii) mid-level in robust herbaceous vegetation or small trees that can support the weight of nest, eggs or young, and the incubating parent (New World blackbirds); (ix) at the top of sturdy vegetation such as trees or snags (Ospreys, certain eagles, herons); (x) old nests in trees (Speckled Teal in Monk Parakeet nests); (xi) tree holes created by woodpeckers (Bufflehead) (Erskine 1971), small tree cavities (Prothonotary Warbler), larger tree cavities or crevices (several sheldgeese, Hooded Mergansers, Wood Ducks), cliff faces, or solid soil banks (kingfishers); and (xii) cliff ledges (several species of geese).

Dependent upon their precocity at hatching and their rate of growth, young of some species like Moorhens (Greij 1994) and coots (Frederickson 1970) use the hatching nest or special brooding ramps over water to provide predator protection. Others like ducks and geese are so precocious that they rarely return to the nest after the first day or two.

4.8 Habitat requirements for molting

Adults of those groups (ducks and geese, coots and rails, grebes, loons, auks, anhingas, flamingos, and dippers) that have a simultaneous or near-simultaneous (as opposed to serial) wing molt annually (some ruddy ducks may do this twice a year) are flightless for 2 to 5 weeks and, therefore, have similar resource needs to those of juveniles: foods and protective water or cover must be available in the same wetland. Prior to this event, many adult waterfowl and probably some other groups undergo a "molt-migration" (Salomonsen 1968), seeking large and protective water bodies that are free of major predators for their flightless period (Hochbaum 1944). Some duck species migrate hundreds of miles to special molting areas after breeding: dabbling ducks (Oring 1968), bay diving ducks (Bergman 1973), Paradise Shelducks (Williams 1979), and some geese (Abraham 1980). Food in abundance also is an obvious necessity during an energy-demanding period. Moreover, some Arctic species migrate first and molt during autumn and winter to avoid freeze-up of northern nesting areas (e.g., eiders (Palmer 1962, Thompson and Person 1963) and loons

(Stresemann and Stresemann 1966, Woolfenden 1967)). Fulvous Whistling Ducks may avoid drying wetlands in late summer by preliminary southward migration prior to the wing molt (Hohman and Richard 1994).

References

Abraham, K. (1980). Moult migration of Lesser Snow Geese. *Wildfowl* 31, 89–93.

Adair, S. E., Moore, J. L., and Kiel, W. H. Jr (1996). Wintering diving duck use of coastal ponds: an analysis of alternative hypotheses. *Journal of Wildlife Management* 60, 83–93.

Bergman, R. D. (1973). Use of southern boreal lakes by postbreeding canvasbacks and redheads. *Journal of Wildlife Management* 37, 160–70.

Bellrose, F. C. and Holm, D. J. (1994). *Ecology and management of the Wood Duck.* Washington, DC: Wildlife Management Institute.

Bildstein, K. (1993). *White Ibis; wetland wanderer.* Washington, DC: Smithsonian Press.

Bradley, E. L. and Holmes, W. N. (1972). The role of nasal glands in the survival of ducks (*Anas platyrhynchos*) exposed to hypertonic saline drinking water. *Canadian Journal of Zoology* 50, 611–17.

Brinson, M. M. (1993). *A hydrogeomorphic classification for wetlands.* Technical Report WRP-DE-4. Vicksburg, MS: US Army Corps of Engineers Waterways Experiment Station.

Brodsky, L. M. and Weatherhead, P. J. (1984). Behavioural thermoregulation in Black Ducks: roosting and resting. *Canadian Journal of Zoology* 62, 1223–6.

Burger, J., Trout, J. R., Wander, W., and Ritter, G. S. (1984). Jamaica Bay studies VII: Factors affecting the distribution and abundance of ducks in a New York estuary. *Estuaries, Coastal and Shelf Science* 18, 673–89.

Bustnes, J. O. and Lonne, O. J. (1995). Sea ducks as predators on sea urchins in a northern kelp forest. In *Ecology of fjords and coastal waters*, eds. H. R. Skjoldal, C. Hopkins, K. E. Erikstad, and H. P. Leinaas, pp. 599–608. Amsterdam: Elsevier.

Cooch, F. G. (1964). A preliminary study of the survival value of a functional salt gland in prairie Anatidae. *Auk* 81, 380–93.

Cornelius, S. E. (1977). Food resource utilization of wintering redheads on Lower Laguna Madre, Texas. *Journal of Wildlife Management* 41, 374–85.

Eldridge, J. L. (1986). Territoriality in a river specialist, the Blue Duck. *Wildfowl* 37, 123–35.

Eldridge, J. L. and Krapu, G. L. (1988). The influence of diet quality on clutch size and laying pattern in mallards. *Auk* 105, 102–10.

Erskine, A. J. (1971). *Buffleheads.* Monograph Series No. 4. Ottawa: Canadian Wildlife Service.

Espino-Barros, R. and Baldassarre, G. A.(1989). Activity and habitat-use patterns of breeding Carribean Flamingos in Yucatan, Mexico. *Condor* 91, 585–91.

Fredrickson, L. H. (1969). Mortality of coots during severe spring weather. *Wilson Bulletin* 81, 450–53.

Fredrickson, L. H. (1970). The breeding biology of American coots in Iowa. *Wilson Bulletin* 82, 445–57.

Greij, E. J. (1994). Common moorhen. In *Migratory shore and upland game bird management in North America*, eds. T. C. Tacha and C. E. Braun, pp. 145–57. Washington DC: International Association of Fish and Wildlife Agencies.

Heitmeyer, M. E. (1987). Protein costs of prebasic molt of female Mallards. *Condor* **90**, 263–6.

Hochbaum, H. A. (1944). *The canvasback on a prairie marsh*. Washington, DC: American Wildlife Institute.

Hohman, W. L. and Richard, D. M. (1994). Timing of remigial molt in Fulvous Whistling Ducks in Louisiana. *Southwestern Naturalist* **39**, 190–2.

Jorde, D. G., Krapu, G. L., Crawford, R. D., and Hay. M. A. (1984). Effects of weather on habitat selection and behavior of mallards wintering in Nebraska. *Condor* **86**, 258–65.

Kear, J. (1972). The Blue Duck of New Zealand. *Living Bird* **11**, 175–92.

Kelly, J. P. and Wood, C. (1996). Diurnal, intraseasonal, and intersexual variation in foraging behavior of the common yellowthroat. *Condor* **98**, 491–500.

Knopf, F. L. and Sedgwick, J. A. (1992). An experimental study of nest-site selection by Yellow Warblers. *Condor* **94**, 734–42.

Krapu, G. L. (1974). Feeding ecology of pintail hens during reproduction. *Auk* **91**, 278–90.

Lack, D. (1968). *Ecological adaptations for breeding in birds*. London: Methuen.

Longcore, J. R. and Gibbs, J. P. (1988). Distribution and numbers of American black ducks along the coast of Maine during the severe weather of 1980–81. In *Waterfowl in winter*, ed. M. W. Weller, pp. 377–89. Minneapolis, MN: University of Minnesota Press.

Marshall, R. M. and Reinert, S. E. (1990). Breeding ecology of seaside sparrows in a Massachusetts salt marsh. *Wilson Bulletin* **102**, 501–13.

McKinney, F., Siegfried, W. R., Ball, I. J., and Frost, P. G. H. (1978). Behavioral specializations for river life in the African black duck (Anas sparsa Eyton). *Zeitschrift für Tierpsychologie* **48**, 349–400.

Miller, M..W. and Nudds, T. D. (1996). Prairie landscape change and flooding in the Mississippi River Valley. *Conservation Biology* **10**, 847–53.

Moulton, D. M. and Weller, M. W. (1984). Biology and conservation of the Laysan Duck (*Anas laysanensis*). *Condor* **86**, 105–17.

Odum, W. E., Smith III, T. J., Hoover, J. K., and McIvor, C. C. (1984). *The ecology of tidal freshwater marshes of the United States east coast: a community profile*. FWS/OBS-83/17. Washington, DC: US Fish and Wildlife Service

Oring, L. W. (1968). Behavior and ecology of certain ducks during the post-breeding period. *Journal of Wildlife Management* **28**, 223–33.

Palmer, R. S. (1962). *Handbook of birds of North America*, Vol. 1. New Haven, CT: Yale University Press.

Petersen, M. R., Douglas, D. C. and Mulcahy, D. M. (1995). Use of implanted satellite transmitters to locate spectacled eiders at sea. *Condor* **97**, 276–8.

Riemer, D. N. (1984). *Introduction to freshwater vegetation*. Westport, CN: AVI Publishing.

Rottenborn, S. C. (1996). The use of coastal agricultural fields in Virginia as foraging habitat by shorebirds. *Wilson Bulletin* **108**, 783–96.

Salomonsen, F. (1968). The moult migration. *Wildfowl* **19**, 5–24.

Schmidt-Nielsen, K. and Kim, Y. T. (1964). The effect of salt intake on the size and function of the salt glands in ducks. *Auk* **81**, 160–72.

Schulthorpe, C. D. (1967). *The biology of aquatic vascular plants.* New York: St Martin's Press,

Siegfried, W. R. (1968). The Black Duck in the South-western Cape. *Ostrich* **39**, 61–75.

Stanley, E. H., Fisher, S. G., and Grimm, N. B. (1997). Ecosystem expansion and contraction in streams. *Bioscience* **47**, 427–35.

Stresemann, E. and Stresemann, V. (1966). Die Mauser der Vogel. *Journal für Ornithologie* Sonderheft **107**, 1–477.

Tacha, T. C., Nesbitt, S. A., and Vohs, P. A. (1994). Sandhill crane. In *Migratory shore and upland game bird management in North America,* eds. T. C. Tacha and C. E. Braun, pp. 77–94. Washington DC: International Association Of Fish and Wildlife Agencies.

Thompson, D. Q. and Person, R. A. (1963). The eider pass at Point Barrow, Alaska. *Journal of Wildlife Management* **27**, 348–56.

Tietje, W. D. and Teer, J. G. (1988). Winter body condition of Northern Shovelers on freshwater and saline habitats. In *Waterfowl in winter,* ed. M. W. Weller, pp. 353–76. Minneapolis, MN: Univ. of Minnesota Press.

Tyler, S. and Ormerod, S. (1994). *The dippers.* London: T and AD Poyser.

Weller, M. W. (1974). Habitat selection and feeding patterns of Brown Teal (*Anas castanea chlorotis*) on Great Barrier Island. *Notornis* **21**, 25–35.

Weller, M. W. (1980). *The island waterfowl.* Ames IA: Iowa State Univ. Press.

Weller, M. W. (1995). Use of two waterbird guilds as evaluation tools for the Kissimmee River restoration. *Restoration Ecology* **3**, 211–24.

Weller, M. W., Adams, I. C., and Rose, B. J. (1955). Winter roosts of marsh hawks and short-eared owls in Central Missouri. *Wilson Bulletin* **67**, 189–93.

Williams, M. (1979). The moult gatherings of Paradise Shelduck in the Gisborne-East District. *Notornis* **26**, 369–90.

Withers, K. and Chapman, B. R. (1993). Seasonal abundance and habitat use of shore-birds on an Oso Bay mudflat, Corpus Christi, Texas. *Journal of Field Ornithology* **64**, 382–92.

Woodin, M. C. (1994). Use of saltwater and freshwater habitats by wintering redheads in southern Texas. In *Hydrobiologia,* Vol. 279–80: *Aquatic birds in the trophic web of lakes,* ed. K. K. Kerekes, pp. 279–87.

Wooley, J. B. Jr and Owen, R. B. (1978). Energy costs of activity and daily energy expenditures in the black duck. *Journal of Wildlife Management* **42**, 739–45.

Woolfenden, G. E. (1967). Selection for a delayed simultaneous wing molt in loons (Gavidae). *Wilson Bulletin* **79**, 416–20.

5

Foods, feeding tactics, strategies, and guilds

Food resources within wetlands can be diverse and vary temporally and spatially. Birds are unique among vertebrates in their ability to use wetlands dispersed over hundreds or thousands of miles in their annual range. However, they are most restricted in movement during the nesting and rearing season or when they are flightless owing to age or annual wing molt. In addition to flight, they may have the flexibility of swimming, running, or walking as needed and efficient, and this reflects the end-products of long years of trial and error. This should not infer that all wetland birds are wandering long distances in search of food, but that they could and sometimes do. Typically, they are bound by a series of historical and evolutionary events that have resulted in the development of successful patterns of movement and breeding strategies for a given region, which is intimately tied to food resources.

Finding accurate information on foods used by birds seasonally and at various life-cycle stages and ages is difficult. An old but still valuable summary of food use by birds in North America is Martin, Zimm and Nelson (1951), but this and papers of this period contain some errors resulting from the examination of gizzards rather than esophagi alone for some bird groups. This underestimates soft animal matter and exaggerates hard seeds. More available but less quantitative summaries of general food uses are in Erlich, Dobkin and Wheye (1988) and Kaufman (1996) in the USA and in similar treaments of bird life histories in each region or country. In addition, there are numerous detailed studies of selected species; however review papers for various taxa, where they exist, are the best source.

5.1 Potential foods typical of wetlands

General categories of food resources used by waterbirds are shown as the top layer in Fig. 5.1 and are products of the diversity of vegetation and animals, which are themselves related to hydroperiods (i.e., duration of water in days, weeks, or months per year), timing of biological and environmental events (e.g., seasonal chronology), and water depths in different wetland types (lower tier in Fig. 5.1). This schematic model reflects my interpretations of physical

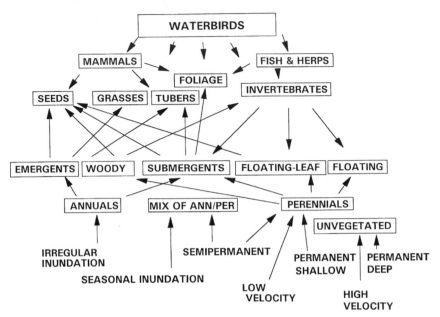

Figure 5.1. Influence of water regimes and velocity on vegetation groups and potential plant foods, and on invertebrate and vertebrate (fish, amphibians and reptiles) foods of waterbirds. (Modified from Weller (1995) Use of two waterbird guilds as evaluation tools for the Kissimmee River restoration. *Restoration Ecology* 3, 211–14. Reprinted by permission of Blackwell Science, Inc.).

forces that drive vegetation. Annual plants are especially common to temporary floodwaters and non-flowing waters; they produce large seed crops useful to pioneering species. Plants of longer hydroperiods tend to produce tubers or rhizomes, but some also produce seed concurrently or under certain stressful conditions such as drought. Vegetation structure is influenced by adaptations to water permanence, timing, current, and climatic and salinity regimes, resulting in microhabitats with differing vegetation structures such as nonpersistent emergents, persistent or robust emergents with sufficient structure that they survive into the next growing season or longer, woody forest or shrub species that may grow continually for many years, or aquatic beds made up of diverse annual or perennial underwater plants that tend to reoccur annually when conditions permit (Fig. 5.1).

Aquatic vertebrates like amphibians and reptiles ("herps"), fish, or wetland mammals like rice rats are major foods of predaceous wetland birds and other vertebrates. But as consumers, they also are linked directly to vegetation, invertebrates, or other smaller vertebrates for food; directly to water for protection; and indirectly to water as the driving force for wetland plant and animal succession over time. Vertebrates of temporary waters, like amphibians, have periodic population bursts based on time-limited water, plant, and invertebrate resources: suddenly they are gone – many eaten by wetland predaceous birds! More aquatic vertebrates like fish, while geared to longer hydroperiods and more permanent areas, include species such as the mosquito fish, which populate, thrive, and die in temporary waters of shallow basins or intermittent streams. These are also eaten by wetland birds when vulnerable. Fish eggs often

Foods, strategies, and guilds

are laid in shallow waters and are a favorite food of many birds, especially in the spring before the egg-laying period.

Although smaller, the mass and diversity of invertebrates makes them the choice food for many breeding birds and for some species all year. Many if not most wetlands are rich in plant biomass and, therefore, are dominated by herbivores (mostly invertebrates but some mammals and birds) and detrivores (mostly invertebrates). A recent summary of data on use of invertebrates by shorebirds showed over 400 genera of 12 phyla, with significant flexibility in use (Skagen and Oman 1996). In some species, birds at stopovers may double their body weight in a few weeks while feeding on invertebrate eggs and larva (Harrington 1996). Invertebrates also are regulated by annual hydroperiods and year-to-year water permanence: short-cycle invertebrates vary dramatically in abundance according to timing and duration of water; long-lived invertebrates with long developmental stages like dragonflies are typical of longer hydroperiods and even more permanent waters. Birds are well adapted to exploit this range of wetland habitats. They move easily, test the resources, and make a choice to establish breeding areas or wintering sites; when they have accomplished their functional goals, or when food resources or water regimes change, they move on.

What foods are rare or absent in wetlands that might influence which bird species can survive? There are few flowering plants among the truly aquatic plants of lake systems (Philbrick and Les 1996), but more probably occur in wetlands. Few, however, have large and prominent flowers except those that float (e.g., water lilies); therefore, it is not surprising that there seem to be few nectar-feeders among wetland birds. Hummingbirds do use jewelweed and other plants of wetland margins that have flowers placed at a suitable and safe height for feeding, but getting nectar from a water lily could be tricky. Moreover, some water lilies have time-regulated, spiraling stems that pull the flower head beneath the water's surface, where the seeds can mature in a protected environment (Riemer 1984). However, diving seed eaters like wintering Ring-necked Ducks tap such underwater resources as a major food in southeastern USA wetlands. There also seem to be few plants that produce large berries and other fleshy fruits in herbaceous and shrub wetlands; as a result, birds like tanagers are not prominent except in forested areas, and they tend to avoid water. Orioles may feed on nectar or fruits, and while they use tall trees adjacent to wetlands, they seem to avoid feeding in wetlands or nesting overwater. Seed production can be impressive, and ground-feeding birds like sparrows, doves and pigeons, quail, and pheasant may use shallow wetlands when they dry in autumn and early winter, but dense vegetation is a danger to larger ground-feeding species.

A model of these habitat/food/bird relationships (Fig. 5.2) may provide some perspective on the dynamics and interrelationships and provide a basis for the topics that follow. How the bird feeds, what it eats, how efficient it is, and how it might compete with other species are strongly influenced by the

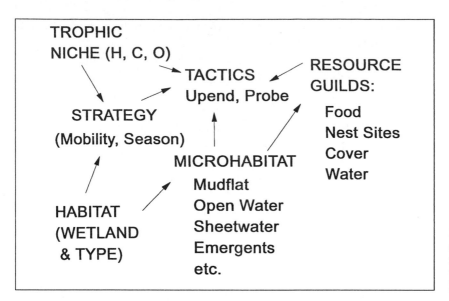

TROPHIC
NICHE (H, C, O)

TACTICS
Upend, Probe

RESOURCE
GUILDS:

Food
Nest Sites
Cover
Water

STRATEGY
(Mobility, Season)

MICROHABITAT
Mudflat
Open Water
Sheetwater
Emergents
etc.

HABITAT
(WETLAND
& TYPE)

Figure 5.2. Potential relationships between trophic or food niche, foraging strategies, tactics, resources guilds, and habitat use at several levels.

nature of the wetland it chooses. Its food choice is based on its trophic niche and its morphological and physiological adaptations, all steeped in evolutionary history. Therefore, wetland choice, and the precise habitat or microhabitat within the wetland, and food needs for specific stages in the annual life cycle influence how the bird forages: its overall strategy and its immediate tactic for food intake (Table 5.1).

5.2 Foraging behavior

Before examining factors that influence food use by waterbirds, it is essential to think in terms of survival needs and adaptations that impose restrictions on how and why birds feed as they do. Most center about energy efficiency in obtaining food. The searching for and intake of food is termed **foraging** and obviously must be efficient as well as effective if the animal is to survive. We often assume that foods are abundant and unlimited for birds, but starvation and poor health is common, and many studies of food habits of birds have noted individuals that have spent long periods foraging but had near-empty digestive tracts when collected. I have seen several cases where waterbirds are limited in access to foods and show intention movements and push the limits to get to food quickly. Kelp Geese of Patagonia and offshore islands feed at low tide by walking among the rocky shore to gather sea lettuce and other large and leafy but often heavily grazed seaweeds. But tides are extreme at that latitude and the wait is often long, and some individuals will arrive at feeding areas and swim above the rocks, even tipping-up like ducks, in an effort to obtain food quickly. Another example of time pressurers is where light limits availability:

Table 5.1 *Common bird feeding sites and foods in wetlands*

Location in wetland	Microhabitat	Examples of major foods
Basin substrate	Epibenthic	Seeds, seedlings, snails, crayfish
	Endobenthic	Roots, tubers, worms, clams
Mudflat	Dry (epibenthic)	Insects, seeds, seedlings
	Wet (endobenthic)	Worms, crabs, clams, crustaceans
Water column	Planktonic (drifters)	Daphnia, cyclops
	Neustonic (surface)	Insects, some duckweeds
	Nekton(underwater swimmers)	Fish and eggs
	Plant leaves and seeds	Filimentous algae, invertebrates
Surface	Plant leaves and seeds	Pondweeds, algae, duckweeds
	Water's surface	Dead or stunned fish and invertebrates
Above water	Stems, bole, leaves	Cattail, bullrush, seeds
Shoreline	Mudflats, bare soil	Seeds, invertebrates
	Herbaceous plants	Grasses, sedges, spikerush
	Herbaceous plant cover	Frogs, snakes, mammals, birds
	Woody plants	Seeds, seedlings, fruits, nuts, berries

White Pelicans in the Gulf of Mexico capture fish that at night are attracted to pier lights. These birds fly from daytime roosts, often arriving before capture is efficient, and loaf until darkness arrives and the lights go on. In both examples, one assumes that hunger triggers the feeding response and that feeding is less successful elsewhere.

The optimization of effort or energy expenditures in the search for food has been termed the **optimal foraging theory** (Stephens and Krebs 1986); it assumes that birds follow the most efficient strategies and tactics that produce the needed food. It helps us to understand why the same species of birds do not all feed in the same way in all areas or at all times. They are doing what it takes in that particular situation to gain the best quality food in the shortest time and with the least effort. This also infers evolutionary adaptations and selection toward certain feeding strategies, and specializations may be the end-product. However, the roles of habit, learning experiences, and innate flexibility are more difficult to evaluate.

5.3. Foraging strategies and tactics

Separating strategies from tactics in patterns of gathering food is relative but perhaps worthwhile because it adds a temporal and evolutionary component (Ellis *et al.* 1976). I use the term strategy to represent a series of approaches or steps to accomplish a major goal incorporating various feeding, breeding, and mobility patterns that ensure security, health, and survival of the species during one or more life-history stages (more in the sense of diet selection, e.g., Ellis *et al.* 1976). Such patterns are not simply a matter of choice by the bird but may be fixed by traditional or genetic adaptations that reflect an evolutionary history which maximizes efficiency and ensures resources in these variable environments. For each species, and probably for some regional populations, the complex of habitat resources required and patterns of mobility to accomplish this result from a long evolutionary history that has pre-adapted the bird for what we consider typical habitat, resulting in preference for and selection of that structural habitat pattern. Hence, there is a scale of foraging dictated by general foods and strategies: a fish-eating cormorant, a duck diving for invertebrates that are hidden in rocks, and a shorebird moving along the shore to opportunistically grab clams or worms exposed by the waves all have different scales and rates of movement as well as different anatomical specializations. These adaptations may be sufficiently unique to create species termed **specialists**, such as the Snail Kite of the Everglades (Sykes 1987). These kites seem limited by their food, Apple Snails, which are characteristic of semipermanent wetlands; this potentially creates a crisis for the species when that habitat is impacted naturally or by human intervention. Other species have been termed **generalists**: feeding on whatever is most available and using diverse nesting and resting sites. Such species have been able to fit into a variety of available habitats, perhaps through exposure to dynamic conditions over areas and time, and may be so widespread and abundant as to be a nuisance (e.g., many New World blackbirds).

Several examples of strategies may indicate the complexity and diversity of approaches required to meet annual needs. (i) Dabbling ducks feed on invertebrates in summer ponds and on seeds or foliage during wintering periods, thus tapping the greatest abundance of each at the peak seasons. Some move to marine habitats, but most favor wintering habitats that are similar to their breeding areas. (ii) Redhead ducks use inland freshwaters for breeding and molting; here, they and their young feed by diving in shallow water for benthic invertebrates. In winter, however, they frequent mainly estuaries or large freshwater lakes, using quite different foods. (iii) Red Phalaropes extend such habitat shifts further by moving from small freshwater tundra ponds to wintering well out at sea where they feed on invertebrates derived from upwelling waters. (iv) The northward migration of Red Knots to tundra for breeding areas is timed to spawning by Horseshoe Crabs on limited beach sites of the Atlantic coast of the USA and to similar invertebrate pulses elsewhere. After breeding, they use the resources of vast tidal mudflats rich in small clams and other marine invertebrates, especially in migration and wintering in the southern hemisphere. (v) Sedge Wrens seem to come and go as a breeding bird of the Prairie Pothole Region and even breed at varying times of the summer dependent on the geographic and seasonal distribution of rainfall. (vi) Wood Storks adjust their breeding times to drying conditions, which create opportu-

Foods, strategies, and guilds

Table 5.2 *Examples of food and foraging tactics employed by some wetland birds*

Guild (as denoted by habitat/locomotion/ food taken)	Foods	Tactics	Species examples
Surface swimmers	Invertebrates and seeds	Strain/sweep/grab	Shoveler, Green-winged Teal
		Dabble/grab	Blue-winged Teal
	Invertebrates	Search/dipping	Phalarope
	Invertebrates and algae	Inverted straining	Flamingos
Water column divers	Fish/invertebrates	Visual search/grab	Grebes, ducks
	Fish	Search/spear	Anhinga
Benthic divers	Invertebrates	Benthic straining	Ruddy Duck, Scaup
Benthic probers	Tubers	Digging/rooting	Snow Goose, cranes, Magpie Goose
Flight-feeders	Insects	Aerial hawking	Franklin's Gull, flycatcher
	Fish	Surface skim	Skimmer
	Carrion	Surface grab	Many gulls
	Fish	Surface grab	Fishing owls and eagles
	Fish/invertebrates	Shallow dive	Terns
	Fish	Plunge-diver	Osprey, Pelican
Perch-divers	Fish/invertebrates	Plunge-diver	Kingfishers
Walkers/runners (out-of-water)	Invertebrates	Search, grab	Sandpipers, plovers
	Insects	Run–snatch	Whimbrel, Laysan Duck
Waders/waiters (in-water)	Fish	Search/strike	Many herons
	Fish/invertebrates	Probe/grab/sweep	Ibises, spoonbills, storks
	Fish/mice	Wait	Great Blue Heron
	Fish/insects	Stalk	Great and Cattle Egrets
	Fish	Wing-flashing	Reddish Egret
	Fish	Wing-shading	Black Heron
	Invertebrates	Probe	Dowitcher
	Minnows	Baiting	Green and Tricolor Herons
Impact feeders	Clams/shellfish	Drop on hard surface	Gulls
	Large eggs	Use of rocks	Bristle-thighed Curlew

nities for access to fish and other prey needed for nestlings. To accomplish this, they shift breeding times and breeding areas.

Tactics and techniques (the latter I think of as more simple that the former) of obtaining food are mechanistic but efficient actions that are still fairly consistent but do vary by food and microhabitat conditions. I have not attempted here to define, differentiate, or describe and name these variations but rather to indicate the array of techniques one can find (Table 5.2). Ashmole (1971) described many techniques used by seabirds that are similar to those of wetland species, and Erlich *et al.* (1988) have added secondary as well as primary techniques for some species – which help to demonstrate how birds may differ by season, location, and opportunity. In no way should one construe these as fixed, as new approaches are developed constantly by birds to meet the changing situation of obtaining food. An interesting technique reflecting good learning ability by the bird has been noted in several species of heron that take bread or other potential fish food from picnickers and place it in the water where the heron can then capture the attracted fish. Several recent observations of kingfishers and grebes feeding on food seemingly made available by feeding jacanas and other waterbirds (Croft 1996), by swimming Platypuses (Burnett 1996, Roberts 1995), and by Southern Otters (McCall 1996) reflect great opportunism. Tool or impact use to open sealed foods is not uncommon in birds, and several waterbirds use these techniques: many gulls drop clams and other shelled invertebrates from the air onto hard surfaces like rocks or rooftops, and Bristle-thighed Curlews have been known to use rocks to open abandoned albatross eggs (Marks and Hall 1992).

Egrets, often with conspicuously colored feet, use foot-stirring to disturb or attract prey, and some plovers and sandpipers engage in foot-stamping, seemingly to make their tiny prey surface. Gadwalls and wigeon, both herbivores, do not dive very efficiently but steal from diving coots as they surface. Piracy practiced by frigatebirds on gulls, by pelicans on cormorants, and by gulls on smaller birds is commonplace. Response of birds to fish spawn is an opportunistic behavior demonstrating dietary needs for high protein in spring and also flexibility in tactics, causing unusual mixed flocks like brant, scoters, goldeneyes, mergansers, and various gulls.

5.4 Guilds

Those species (related or not) that exploit the same resources in a similar way can be termed guilds. Although this term is not used consistently, ecologists have found it useful in describing important resource-seeking attributes when comparing different species of birds or animals as diverse as birds and bees. Some of the terms in common use for various guilds and others that I have coined are listed in Table 5.2; these are merely examples and not intended as a descriptive system. It may be, in fact, better to maintain flexibility in terms but to structure these hierarchically, to allow comparison at scales from large to small. Feeding sites or microhabitats (Table 5.1), i.e., the physical position in the wetland, induce physical limitations on food gathering and thus influence behavioral tactics and the use of locomotion, or the design of bill, leg, and sometimes neck.

In attempting to apply information on guilds to assess characteristics of a wetland or bird community, a common mistake is to use a taxonomic category such as waterfowl, assuming that all species in the group are similar in foods

Foods, strategies, and guilds

because they have generally similar body shapes, bills, feet, etc. In fact, there are over 150 species of waterfowl worldwide (Table 3.1) and they range from pure herbivores that walk on land most of the time to pure carnivores that eat animal foods all year. Most are omnivores, and their foods differ by season, so temporal scale as well as taxonomic level is vital in making realistic comparisons. Even comparing dabbling ducks with bay ducks is not a safe approach because each tribe has a range of foods, microsites, and diving depths that segregate the species (e.g., Smith *et al.* 1986). For example, three dabbling ducks are commonly seen in open water (shoveler, wigeon, and Gadwall) along with coots, whereas most other dabbling ducks favor vegetated or otherwise protected areas. Shovelers typically sweep their bills side-to-side while straining for plankton whereas wigeon and Gadwall are grabbing and cutting leaves while feeding bill-under or upending. The coots (as well as bay ducks) may dive to grab submergent vegetation such as wigeongrass or pondweeds, which then is pirated by the widgeon.

At some times and places, shovelers may be followed by phalaropes, which are not filter-feeders but are tapping the same invertebrates by their unique probing action, sometimes swimming in circles (hence the name "whirligig," after the water beetle). The shovelers apparently are acting as "beaters" by stirring up invertebrates by their swimming actions. Shovelers use the same tactic following swans, which are seeking submergent vegetation rather than invertebrates.

Another example showing how foods might be misleading in classifying guilds is that of the Virginia Rail and Blue-winged Teal. Both are often omnivores during summer and both might have snails and seeds in their digestive tract. However, because their feeding techniques (walking versus swimming) and microsites (tall and dense vegetation versus semiopen, shallow water) differ, most biologists would not consider them in the same guild, if categorizing at a fine scale. Some workers have noted significant variation in foraging sites seasonally or through the day (Kelly and Wood 1996), part of which could be a result of targeting available foods or selecting foods on the basis of needed nutrition or energy. Therefore, limited observations at one time could be misleading and one must recognize such potential influences.

5.5. The where, when, and why of food use

The importance of the timing of food is that maintenance needs must be met before energy can be directed to flight, migration, breeding, defense, etc. High-protein food for laying females or growing young is crucial, and birds are adept at searching for it. Presumably, pre-nesting females would not settle in a nesting area unless such food is present, but some species have strategies that ensure resources arrive with them in the form of body fat and tissue. Pre-nesting fattening is common among waterfowl (Alisaukas and Ankney 1992,

Krapu and Reineke 1992) and coots, but the degree to which this is necessary for laying seems to vary by species (Arnold and Ankney 1997) and probably habitat conditions at the site. During spring migration, most omnivorous dabbling ducks seem to fatten and also to increase their intake of invertebrates prior to arrival at breeding marshes and, therefore, are less dependent on local food conditions; those that nest later may fatten in the breeding marsh on local foods (Northern Shovelers), and South African Red-billed Teal seemingly are able to lay with minimal intake of invertebrates (Petrie 1996).

During nesting, the precocity of the young determines where foods are obtained: colonial nesters like egrets, herons, pelicans, cormorants (see review in Weller 1995) carry food many miles to feed young in the nest. Coots, rails, and grebes swim within a few feet of the nest until the young are able to join them. Young grebes seem more inclined than most swimmers to ride on the back of adults for longer water travel. Although some ducklings can walk several miles (Hochbaum 1944) and swim strong river currents when newly hatched (Bellrose and Holm 1994), it is an obvious efficiency to have abundant foods nearby.

Protein-rich feeding areas also are important for birds that shed their flight feathers simultaneously, and it is not surprising that post-breeding molt-migrations of waterfowl take place at a time for that species that ensures both food and water. Most ducks do this in mid to late summer, and movement to more permanent waters is common, some even flying northward to ensure protection from predators when small water areas might dry up (Bailey and Titman 1984, Bergman 1973). Loons (Palmer 1962) and some eiders delay wing molt until reaching wintering or at least fall staging areas (Hohman, Ankney and Gordon 1992). After molt, migrant and wintering birds are more likely to use high-energy sources like carbohydrates (seeds, foliage, fruits) to maintain body condition and sustain normal activities (Weller 1975).

Body "condition," as judged by weight and fat storage, is a major requirement for migration in many bird species, especially those that fly long distances between stops (e.g., Ruddy Turnstones and Red Knots, Harrington 1996). It is fairly common for a few shorebirds to remain in wintering areas all year, presumably because they are diseased, injured, or their body condition has not reached a level essential for the long flight to Arctic areas (Johnston and McFarlane 1967). Some short-range migrants do not seem to show this need to fatten on long but few stopovers (Skagen and Oman 1996), but undoubtedly many factors influence this strategy.

Many factors influence where foods are found within a wetland (i.e., spatial distribution): bottom substrate materials, vegetation at all levels (layers), water depth, tidal regimes, water temperature, or oxygen supply in the water. For example, bottom sediments of sand might be suitable for clams and not for midges. Submerged vegetation provides substrates for countless invertebrates of varying sizes and food requirements. Emerging vegetation such as cattail provides insect emergence sites for organisms like damselflies and dragonflies

Foods, strategies, and guilds

that live underwater for months or years and crawl up vegetation to dry and emerge from their juvenile exoskeleton as flying adults. Dead and dying vegetation is suitable for detritivores when herbivores would not flourish, but detritivores typically are dominant organisms in energy cycles of wetlands and, therefore, are likely to be major food items for many species of bird. Water depth influences temperature and oxygen levels in water, and thereby the fish or invertebrate species than can thrive there. Declining water depth and low oxygen levels make invertebrates and fish vulnerable to predation by ibises, egrets, and herons. Mudflats are rich places for many invertebrates, and shore-birds focus their search there. These microsites strongly influence where birds are found in a wetland. Table 5.2 provides a partial list of such feeding locations within the wetland, more specific feeding sites, food types, and examples of fairly common bird foods.

Foods used by birds vary with their body size, and although many have bills and tactics designed for essential flexibility, they can be more efficient by taking large prey (if of equal quality) and spending less energy chasing many small items. For example, large herons tend to eat larger fish than small herons (see Weller 1995 for review); large ducks take larger maximum animal prey than small ones (Pehrsson 1984, Weller 1972). Curlews take larger clams at greater depths than do oystercatchers (Zwarts and Wanink 1984). (Fig. 5.3), which take larger and older clams than do Sanderlings (Myers, Williams and Pitelka 1980). In fact, oystercatchers do not waste time with small clams, and big ones are too deep to reach. Hence, there is an impact on one age-class of clams, and, reciprocally, if the suitable age-class were unavailable, this would decrease habitat (food) quality. This sort of optimization in bird/food size is closely related to specialization, because a suitable bill may increase efficiency of taking prey but decrease flexibility in food options, and may avoid extensive overlap in foods and often habitat. Even specialists like shoveler ducks and spoonbills can strain for multiple small items like zooplankton through the specially designed straining lamellae along the sides of the mandible, or use the nail on the tip of the bill to grab large items like snails when they are conspicu-ously abundant in late summer. In tidal areas, prey availability varies with tide level, and "tide-chasers" like Sanderlings feed where and when food resources are most available and efficiently taken (Connors et al. 1981).

Vulnerability of prey may be affected by numerous factors, such as lighting, underwater plant cover, distraction behavior, or speed of the chase. Piscivores like the Reddish Egret of North America and the Black Heron of Africa use their wings to shade the waters (although in quite different ways), and several species of smaller egrets and the Black Herons have yellow feet, which are thought to attract fish. Black-crowned and Yellow-crowned Night Herons commonly feed at dusk or night, seemingly finding crayfish and crabs espe-cially vulnerable then. Night feeding by ducks has been associated with preda-tor avoidance (Tamisier 1985) and human disturbance (Thompson 1973, Thornburg, 1973), but such foraging is limited to strainer-feeders like shovelers

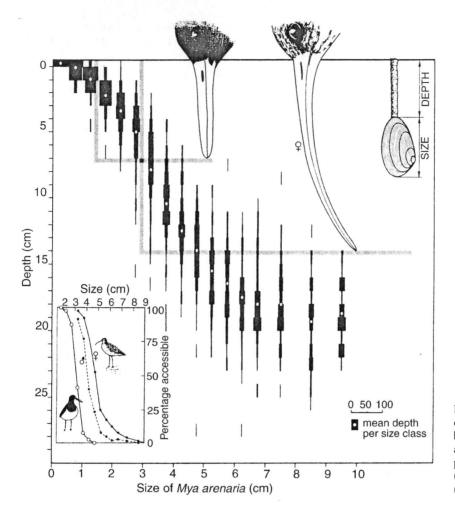

Figure 5.3. Mean depth per clam size-class in relation to bill length of oystercatchers and curlews, which influences potential availability (From Zwarts and Wanink (1984) with permission).

or Green-winged Teal, bottom dabblers or divers that need less visual information to obtain food, and probing mudflat or shallow-water birds like dowitchers, Willets, and curlews. Herons and cormorants seem to be more successful in catching rough fish that are slow-moving.

Use of certain habitats that are potentially more stressful than others suggests the long-term evolution of suitable foraging strategies and adaptations in morphology and physiology for specific foods, which make the bird successful in this environmental setting. For example, most if not all waterbirds have salt glands, but they vary by age and species in how large and efficient they are. Some birds (e.g., Redheads) breed in freshwater probably because the young cannot tolerate the salt, but adults and immatures later switch to higher salinity water in wintering areas; their salt glands change accordingly. Other species are still better adapted and can tolerate salty conditions all year, where they not

Foods, strategies, and guilds

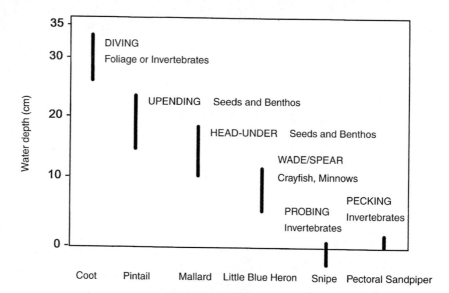

Figure 5.4. Tactics and possible food resources of some common waterbirds in relation to the water depths in which they were recorded (modified from Fredrickson and Taylor 1982).

only survive but reproduce (e.g., Common Eiders, avocets, stilts). Clearly, those species less-well adapted avoid such habitats.

5.6 Habitat segregation as related to food use

A major concern in vertebrate ecology is habitat or resource segregation, possibly resulting in reduced competition for food or increased feeding efficiency. Although we visualize and measure these differences by use of different habitats within a wetland or among different wetland types, food is a major influence. Obviously, this is simplified by major morphological adaptations for feeding (leg length, bill length and shape, and food type); however, many habitat features are intertwined on a more proximate basis. Depth of water when food resources are on the bottom also influences vulnerability of the prey and predator efficiency, thereby determining the assemblage of birds at the feeding site (Fig. 5.4). Large and skilled diving birds like loons, cormorants, and mergansers feed in open water of sufficient depth for them to pursue fish or invertebrates that find such waters suitable. Smaller species like grebes and some ducks take smaller prey and seemingly are adept at navigating around and through duck vegetation. Other species are near-surface or bottom feeders in shallow water (Table 5.1) and adaptations are designed for these depths and conditions (Table 5.2). Common species often segregate feeding efforts by depth of water or substrate and thus reduce competition for foods, as well as capitalize on the efficiency of their feeding adaptations (Fig. 5.3). Some species of closely matched birds (in size or taxonomy or both) may use different foods or foods of different sizes, and they may even have bills to accomplish the

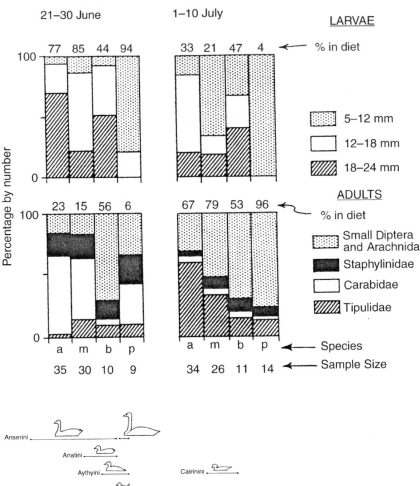

21–30 June   1–10 July   <u>LARVAE</u>

77 85 44 94   33 21 47 4   ← % in diet

Percentage by number

5–12 mm

12–18 mm

18–24 mm

<u>ADULTS</u>

23 15 56 6   67 79 53 96   ← % in diet

Small Diptera and Arachnida

Staphylinidae

Carabidae

Tipulidae

a m b p   a m b p   ← Species

35 30 10 9   34 26 11 14   ← Sample Size

Figure 5.5. Prey size of four species of *Calidris* sandpipers, which decline in bill size from left to right: C. *alpinac*(a), *melanotos*(m), *bairdii*(b), and *pusillus*(p) (From Holmes and Pitelka 1968, with permission).

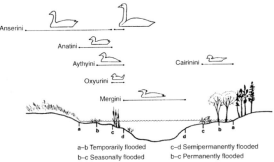

Anserini

Anatini

Aythyini   Cairinini

Oxyurini

Mergini

a–b Temporarily flooded   c–d Semipermanently flooded
b–c Seasonally flooded   b–c Permanently flooded

Figure 5.6. Microhabitats produced by water depth and plant communities as they influence habitat use by various tribes of waterfowl (Anatidae) in prairie wetlands (Krapu and Reineke 1992).

segregation (e.g., Holmes and Pitelka 1968), (Fig. 5.5). Other birds use the same foods in the same wetland but not always at the same time (Connors, Myers and Pitelka 1979) or at the same habitat site (Holmes and Pitelka 1968). In a taxonomic array, adaptation of various taxa demonstrates reduced overlap in foraging sites and foods, as shown in waterfowl that are dominantly swim-feeders (Krapu and Reineke 1992) (Fig. 5.6). For perch-feeders like kingfishers, feeding sites are important in segregation, as shown in tropical species, which may avoid competition by selecting different perch heights from which to fish (Fig. 5.7).

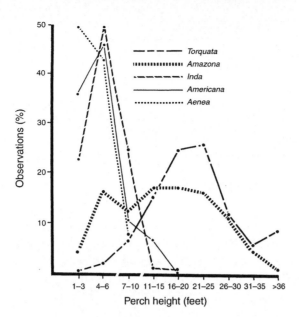

Figure 5.7. Perch heights of five species of neotropical kingfishers (modified from Remsen 1990).

Forster's and Black Terns both nest in prairie potholes but seem segregated by nest sites and cover-water conditions, as well as by foods. They may nest close together in the same wetland, but the Forster's Tern favors relatively elevated structures like muskrat lodges near large and open water where fish are prevalent. In contrast, Black Terns usually feed on insects and are solitary nesters in small pools surrounded by emergent vegetation. They use mats of floating vegetation or muskrat feeding ramps for nests where available.

Some workers feel that the great abundance of foods in some wetland sites and their availability have allowed some closely related species to feed on the same resources at the same time – as with mixed flocks of herons and egrets catching fish or crayfish stranded in pools caused by declining water, macroinvertebrates at their population peak sought by many species of migrant ducks and shorebirds, and feeding by many species at all habitat layers exploiting full grown and emerging marsh flies (Chironomidae) in northern marshes. Water levels, near-freezing water, hot water with reduced oxygen levels, which all influence vulnerability of invertebrates, frogs, and fish, and plant-cover conditions are important in making some resources available. As a result, interpreting species richness or predicting use by different species requires an evaluation of birds in the area physically and behaviorally adapted to efficiently using those foods under the prescribed conditions.

There are still further influences of foods on the presence of waterbirds that will be detailed below: food density, influence on breeding seasons, and impacts of birds on their resources. An assemblage of wetland birds is not a fixed entity, but rather a collection of species with common adaptations that are commonly seen together under certain water or weather conditions. They

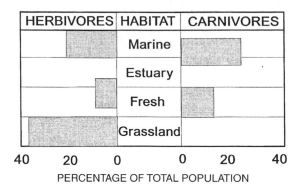

Figure 5.8. Trophic niches and habitat of resident waterfowl of the Falkland Islands, with proportion of the population based on survey data (total = 100%) (modified from Weller 1972 in *Wildfowl* and Weller 1980, with permissions).

are limited by food resources, and, therefore, we expect to see a gross relationship between food types, abundance, and populations of each species or guild. Data pooled on Falkland Island waterfowl (Fig. 5.8) schematically show the relationship of abundance to food or trophic niches and to habitat.

References

Alisauskas, R. T. and Ankney, C. D. (1992). The cost of egg laying and its relationship to nutrient reserves in waterfowl. In *Ecology and management of breeding waterfowl*, eds. B. D. J. Batt, A. D. Afton, M. G. Anderson, C. D. Ankney, D. H. Johnson, J. A. Kadlec, and G. L. Krapu, pp. 30–61. Minneapolis, MN: University of Minnesota Press.

Arnold, T. W. and Ankney, C. D. (1997). The adaptive significance of nutrient reserves to breeding American Coots: a reassessment. *Condor* 99, 91–103.

Ashmole, N. P. (1971). Seabird ecology and the marine environment. In *Avian Biology*, Vol. I, eds. D. S. Farner and J. R. King, pp. 223–86. New York: Academic Press.

Bailey, R. O. and Titman, R. D. (1984). Habitat use and feeding ecology of post-breeding Redheads. *Journal of Wildlife Management* 48, 1144–55.

Bellrose, F. C. and Holm, D. J. (1994). *Ecology and management of the wood duck*. Washington, DC: Wildlife Management Institute.

Bergman, R. D. (1973). Use of southern boreal lakes by postbreeding canvasbacks and redheads. *Journal of Wildlife Management* 37, 160–70.

Burnett, J. (1996). A further observation of a feeding association between the Platypus and the Azure Kingfisher and a discussion of feeding associations between birds and mammals. *Sunbird* 26, 76–8.

Connors, P. G., Myers, J. P., Connors, C. S. W., and Pitelka, F. A. (1981). Interhabitat movements by sanderlings in relation to foraging profitability and the tidal cycle. *Auk* 98, 49–64.

Connors, P. G., Myers, J. P. and Pitelka, F. A. (1979). Seasonal habitat use by Arctic Alaskan shorebirds. *Studies in Avian Biology* 2, 101–11.

Croft, P. (1996). Grebe foraging with jacanas. *Australian Birds* 30, 21–2.

Ellis, J. E., Weins, J. A., Rodell, C. F., and Anway, J. C. (1976). A conceptual model of diet selection as an ecosystem process. *Journal of Theoretical Biology* 60, 93–108.

Erlich, P. R., Dobkin, D. S., and Wheye, D. (1988). *The birder's handbook*. New York: Simon and Schuster.

Fredrickson, L. H. and Taylor, T. S. (1982). *Management of seasonally flooded impoundments for wildlife.* Resource Publication 148. Washington, DC: US Fish & Wildlife Service.

Harrington, B. (1996). *The flight of the Red Knot.* New York: W. W. Norton.

Hochbaum, H. A. (1944). *The Canvasback on a prairie marsh.* Washington, DC: American Wildlife Institute.

Hohman, W. L., Ankney, C. D., and Gordon, D. H. (1992). Ecology and management of postbreeding waterfowl. In *Ecology and management of breeding waterfowl*, eds. B. D. J. Batt, A. D. Afton, M. G. Anderson, C. D. Ankney, D. H. Johnson, J. A. Kadlec, and G. L. Krapu, pp.128–89. Minneapolis, MN: University of Minnesota Press.

Holmes, R. T. and Pitelka, F. A. (1968). Food overlap among coexisting sandpipers on northern Alaska tundra. *Systematic Zoology* 17, 305–18.

Johnston, D. W. and McFarlane, R. W. (1967). Migration and bioenergetics of flights in the Pacific Golden Plover. *Condor* 69, 156–68.

Kaufman, K. (1996). *Lives of North American birds.* New York: Houghton Mifflin.

Kelly J. P. and Wood, C. (1996). Diurnal, intraseasonal, and intersexual variation in foraging behavior of the common yellowthroat. *Condor* 98, 491–500.

Krapu, G. L. and Reinecke, K. J. (1992). Foraging ecology and nutrition. In *Ecology and management of breeding waterfowl*, eds. B. D. J. Batt, A. D. Afton, M. G. Anderson, C. D. Ankney, D. H. Johnson, J. A. Kadlec, and G. L. Krapu, pp. 1–29. Minneapolis, MN: University of Minnesota Press.

Marks, J. S., and Hall, C. S. (1992). Tool use by Bristle-thighed Curlews feeding on albatross eggs. *Condor* 94, 1032–34.

Martin, A. C., Zimm, H. S., and Nelson, A. L. (1951). *American wildlife and plants; a guide to wildlife food habits.* New York: McGraw-Hill.

McCall, R. (1996). A novel foraging association between southern river otter *Lutra longicaudis* and Great Egrets *Casmerodius albus. Bulletin of the British Ornithological Club* 116, 199–200.

Myers, J. P., Williams, S. L., and Pitelka, F. A. (1980). An experimental analysis of prey availability for Sanderlings. (Aves: Scolopacidae) feeding on sandy beach crustaceans. *Canadian Journal of Zoology* 58, 1564–74.

Palmer, R. S. (1962). *Handbook of birds of North America*, Vol. 1. New Haven, CT: Yale University Press.

Pehrsson, O. (1984). Diving duck populations in relation to their food supplies. In *Coastal waders and wildfowl in winter,* eds. P. R. Evans, J. D. Goss-Custard, & W. G. Hale, pp. 101–16. Cambridge: Cambridge University Press.

Petrie, S. A. (1996). Red-billed teal foods in semiarid South Africa: a north temperate contrast. *Journal of Wildlife Management* 60, 874–881.

Philbrick, C. T. and Les, D. H. (1996). Evolution of aquatic angiosperm reproductive systems. *BioScience* 46, 813–26.

Remsen, J. V. Jr (1990). Community ecology of Neotropical Kingfishers. *University of California Publications in Zoology* 124, 1–116.

Riemer, D. N. (1984). *Introduction to freshwater vegetation.* Westport, CT: AVI Publishing.

Roberts, P. (1995). Grebes foraging with platypus. *Australian Birds* 28, 78–9.

Schreiber, R. W. and Ashmole, N. P. (1970). Sea-bird breeding seasons on Christmas Island, Pacific Ocean. *Ibis* 112, 363–94.

Skagen, S. K. and Oman, H. D. (1996). Dietary flexibility of shorebirds in the Western Hemisphere. *Canadian Field-Naturalist* 110, 419–44.

Smith, L. M., Vangilder, L. D., Hoppe, R. T., Morreale, S. J., and Brisbin I. L. Jr (1986). Effects of diving ducks on benthic food resources during winter in South Carolina, USA *Wildfowl* 37, 136–41.

Stephens, D. W. and Krebs, J. R. (1986). *Foraging theory*. Princeton, NJ: Princeton University Press.

Sykes, P. W. Jr (1987). The feeding habits of the Snail Kite in Florida, USA. *Colonial Waterbirds* 10, 84–92.

Tamisier, A. (1985). Some considerations on the social requirements of ducks in winter. *Wildfowl* 36, 104–8.

Thompson, D. (1973). Feeding ecology of diving ducks on Keokuk Pool, Mississippi River. *Journal of Wildlife Management* 37, 367–81.

Thornburg, D. D. (1973). Diving duck movements on Keokuk Pool, Mississippi River. *Journal of Wildlife Management* 37, 382–89.

Weller, M. W. (1972). Ecological studies of Falkland Islands' waterfowl. *Wildfowl* 23, 25–44.

Weller, M. W. (1975). Migratory waterfowl: a hemispheric perspective. In *Symposium on wildlife and its environment in the America's. Universitad Autonoma de Nuevo Leon Publiciones Biologicas Instituto de Investigaciones Cienificas,* 1, 89–130.

Weller, M. W. (1980). *The island waterfowl.* Ames IA: Iowa State University Press.

Weller, M. W. (1995). Use of two waterbird guilds as evaluation tools for the Kissimmee River restoration. *Restoration Ecology* 3, 211–24.

Zwarts, L. and Wanink, J. (1984). How Oystercatchers and Curlews successively deplete clams. In *Coastal waders and wildfowl in winter,* eds. P. R. Evans, J. D. Goss-Custard, & W. G. Hale, pp. 69–83. Cambridge: Cambridge University Press.

Further reading

Bartonek, J. C. and Hickey, J. J. (1969). Food habits of canvasbacks, redheads, and lesser scaup in Manitoba. *Condor* 71, 280–90.

Bengtson, S. A. (1971). Variations in clutch-size in ducks in relation to the food supply. *Ibis* 113, 523–6.

Bossenmaier, E. F. and Marshall, W. H. (1958). Field-feeding by waterfowl in south-western Manitoba. *Wildlife Monograph* 1, 1–32.

Bryant, D. M. (1979). Effects of prey density and site character on estuary usage by overwintering waders (Charadrii). *Estuary & Coastal Marine Science* 9, 369–84.

Chavez-Ramirez, F., Hunt, H. E., Slack, R. D., and Stehn, T. V. (1996). Ecological correlates of whooping crane use of fire-treated upland habitats. *Conservation Biology* 10, 217–23.

Danell, K. and Sjoberg, K. (1977). Seasonal emergence of chironomids in relation to egg laying and hatching of ducks in a restored lake. *Wildfowl* 28, 129–35.

DeGraaf, R. M., Tilghman, N. G., and Anderson, S. H. (1985). Foraging guilds of North American birds. *Environmental Management* 9, 493–536.

Douthwaite, R. J. (1977). Filter-feeding ducks of the Kafue Flats, Zambia, 1971–1973. *Ibis* 119, 44–66.

Douthwaite, R. J. (1978). Geese and red-knobbed coot on the Kafue Flats in Zambia, 1970–1974. *East African Wildlife Journal* 16, 29–47.

Douthwaite, R. J. (1980). Seasonal changes in the food supply, numbers and male plumages of pigmy geese on the Thamalakane river in northern Botswana. *Wildfowl* 31, 94–8.

Emlen, S.T. and Ambrose III, H. W. (1970). Feeding interactions of Snowy Egrets and Red-breasted Mergansers. *Auk* 87, 164–5.

Eriksson, M. O. G. (1978). Lake selection by Goldeneye ducklings in relation to the abundance of food. *Wildfowl* 29, 81–5.

Esler, D. (1990). Avian community responses to hydrilla invasion. *Wilson Bulletin* 102, 427–40.

Evans, P. R. and Dugan, P. J. (1984). Coastal birds: numbers in relation to food resources. In *Coastal waders and wildfowl in winter,* eds. P. R. Evans, J. D. Goss-Custard, & W. G. Hale, pp. 8–28. Cambridge: Cambridge University Press.

Evans, P. R., Goss-Custard, J. D., and Hale, W. G. (eds.) (1984). *Coastal waders and wildfowl in winter.* Cambridge: Cambridge University Press.

Glazener, W.C. (1946). Food habits of wild geese on the Gulf Coast of Texas. *Journal of Wildlife Management* 10, 322–9.

Goodman, D. G. and Fisher, H. I. (1962). *Functional anatomy of the feeding apparatus in waterfowl.* Carbondale, IL: Southern Illinois University Press.

Gray, L. J. (1993). Response of insectivorous birds to emerging aquatic insects in riparian habitats of a tallgrass prairie stream. *American Midland Naturalist* 129, 288–300.

Joyner, D. E. (1980). Influence of invertebrates on pond selection by ducks in Ontario. *Journal of Wildlife Management* 44, 700–5.

Kaminski, R. M., and Prince, H. H. (1981). Dabbling duck activity and foraging response to aquatic macroinvertebrates. *Auk* 98, 115–26.

Kaminski, R. M. and Prince, H. H. (1984). Dabbling duck-habitat associations during spring in the Delta Marsh, Manitoba. *Journal of Wildlife Management* 48, 37–50.

Kantrud, H., Krapu, G. L., and Swanson, G. A. (1989). *Prairie basin wetlands of the Dakotas: a community profile.* Biology Report 85(7.28). Washington, DC: US Fish & Wildlife Service.

Krapu, G. L. (1974). Feeding ecology of pintail hens during reproduction. *Auk* 91, 278–90.

Kushlan, J. A. (1974). Quantitative sampling of fish populations in shallow, freshwater environments. *Transactions of the American Fisheries Society* 103: 348–52.

Lingle, G. R. and Sloan, N. F. (1980). Food habits of White Pelicans during 1976 and 1977 at Chase Lake National Wildlife Refuge, North Dakota. *Wilson Bulletin* 92, 123–25.

Maltby, E. (1986). *Waterlogged wealth.* London and Washington, DC: International Institute for Environment and Development, Earthscan.

Murkin, H. R., Kaminski, R. M., and Titman, R. D. (1982). Responses by dabbling ducks and aquatic invertebrates to an experimentally manipulated cattail marsh. *Canadian Journal of Zoology* 60, 2324–32.

Nelson, J. W. and Kadlec, J. A. (1984). A conceptual approach to relating habitat structure and macroinvertebrate production in freshwater wetlands. *Transactions of the North American Wildlife and Natural Resource Conference* 49, 262–70.

Orians, G. H. (1980). *Some adaptations of marsh-nesting blackbirds.* Princeton, NJ: Princeton University Press.

Orians, G. H. and Horn, H. S. (1969). Overlap in foods and foraging of four species of blackbirds in the Potholes of central Washington. *Ecology* **50**, 930–8.

Pehrsson, O. (1976). Food and feeding grounds of the goldeneye Buchephala clangula (L.) on the Swedish west coast. *Ornis Scandinavia* **7**, 91–112.

Remsen, J. V. Jr and Robinson, S. K. (1990). A classification scheme for foraging behavior of birds in terrestrial habitats. *Studies in Avian Biology* **13**, 144–60.

Root, R. B. (1967). The niche exploitation pattern of the blue-gray gnatcatcher. *Ecological Monographs* **37**, 317–50.

Rosine, W. N. (1955). The distribution of invertebrates on submerged aquatic plant surfaces in Muskee Lake, Colorado. *Ecology* **36**, 308–14.

Rundel, W. D. and Fredrickson, L. H. (1981). Managing seasonally flooded impoundments for migrant rails and shorebirds. *Wildlife Society Bulletin* **9**, 80–7.

Shearer, L. A., Jahn, B. J., and Ienz, L. (1969). Deterioration of duck foods when flooded. *Journal of Wildlife Management* **33**, 1012–15.

Shull, G. H. (1914). The longevity of submerged seed. *Plant World* **17**: 329–37.

Swanson, G. A. and Meyer, M. I. (1973). The role of invertebrates in the feeding ecology of Anatidae during the breeding season. In *The Waterfowl Habitat Management Symposium*, pp. 143–85. Moncton, New Brunswick: The Wildlife Society.

Swanson, G. A. and Meyer, M. I. (1977). Impact of fluctuating water levels on feeding ecology of breeding blue-winged teal. *Journal of Wildlife Management* **41**, 426–33.

Telfair II, R. C. (1983). *The cattle egret: a Texas focus and a world view*. College Station, TX: Texas A&M University Press.

Thayer, G. W., Wolfe, D. A., and Williams, R. B. (1975). The impact of man on seagrass systems. *American Scientist* **63**, 288–96.

Tietje, W. D. and Teer, J. G. (1996). Winter feeding ecology of Northern Shovelers on freshwater and saline wetlands in South Texas. *Journal of Wildlife Management* **60**, 843–55.

Voigts, D. K. (1973). Food niche overlap of two Iowa marsh icterids. *Condor* **75**, 392–9.

Voigts, D. K. (1976). Aquatic invertebrate abundance in relation to changing marsh vegetation. *American Midland Naturalist* **95**, 313–22.

van Eerden, M. R. (1984). Waterfowl movements in relation to food stocks. In *Coastal waders and wildfowl in winter*, eds. P. R. Evans, J. D. Goss-Custard, & W. G. Hale, pp. 84–100. Cambridge: Cambridge University Press.

6

Bird mobility and wetland predictability

Various kinds of bird movement have been mentioned in relation to tracking and exploiting suitable foods and other resources, and here I want to examine the role of wetlands and their resources as forces in the evolution of long-term patterns of mobility, and their relationships to functions essential to the life cycle. There has been considerable concern recently about the requirements of terrestrial neotropical birds that favor large and contiguous habitats in migration, breeding, and wintering. In contrast, we tend to think of wetland birds as flexible users of smaller patches or "habitat-islands," which enables a shift to seasonally better or more predictable wetlands. One should not, however, assume that any wet area they encounter is suitable. It is obvious that geographic targets for many wetland birds are quite small, and that flights over uninhabitable areas (e.g., forest, deserts, or oceans) to reach a small area may be quite perilous. It is known from banded and radio-tagged birds that birds are precise in reaching stopovers or breeding sites, even when moving from or to relatively small land masses like islands (Johnston *et al.* 1997), but what are the chances that the wetland "island" may be too dry – or too wet – and, therefore, uninhabitable?

We also assume that coastal birds have continuous habitat, but, while more hospitable than a forest or desert, many coastal areas are sand, gravel, or rocky cliff suitable for only a few wetland species. Unfortunately, suitable and reliable feeding, resting, nesting, or wintering areas along the coasts tend to be the same ideal habitats for humans – freshwater inflow areas that create rich estuarine wetland complexes. As a result, direct conflict with human development and recreation is inevitable.

The functional requirements of the annual life cycle that often involve and may induce mobility include: (i) meeting individual body maintenance requirements despite seasonal climate changes and physiological needs; (ii) annual migrations between breeding and molting or wintering areas, if they are separated; (iii) obtaining essential nutrients for egg laying; (iv) foraging for food for young on breeding areas;(v) seeking isolation and protection during post-breeding molting periods (especially for those groups that become flightless); and (vi) locating areas rich in maintenance foods during migration and wintering. Some workers have pointed out that wetland variability may

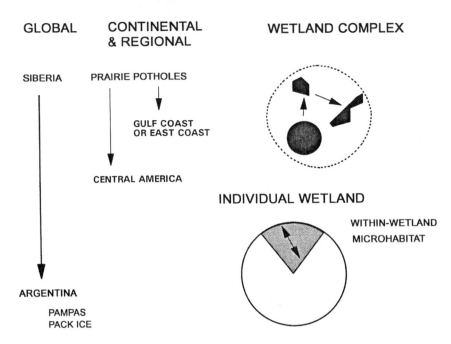

GLOBAL   CONTINENTAL   WETLAND COMPLEX
               & REGIONAL

SIBERIA   PRAIRIE POTHOLES

                 GULF COAST
                 OR EAST COAST

            CENTRAL AMERICA

INDIVIDUAL WETLAND

                       WITHIN-WETLAND
                       MICROHABITAT

ARGENTINA

    PAMPAS
    PACK ICE

Figure 6.1. Examples of the scales of mobility used by wetland birds in exploiting variable resources, reflecting local movements in single wetlands or complexes and regional to global strategies in widely separated wetlands.

induce different strategies at different stages of reproduction (Patterson 1976, Weller 1975). What are the characteristics of wetland birds and the habitats they use that might provide clues as to the origins of different mobility patterns related to geography and function? Examples of some of these patterns have been mentioned earlier, and a few more are added here to focus on wetland influences and possible origins. I will focus on dependability (and hence predictability) of climate and wetland resources, which seems like an obvious key factor to the development of these patterns. This recognizes that orientation capabilities are remarkable and flight amazingly efficient in energy use but also that food and freedom from disturbance are essential at rest stops as well as at destinations. The scale of movements differs by species, group, and function from global to local (Fig. 6.1). We should not assume that there is a single cause for movement by any species nor that all populations of the same species behave similarly. Based on the regular migrations of grebes and rails to even remote places, apparent flight capability may be less important than many would think.

6.1 Latitude and temperature

Latitude and temperature are two influences that are difficult to separate and are complicated by oceanic and altitudinal differences. While best known in the northern hemisphere (Gauthreaux 1982, Lincoln 1935), where climatic regimes make seasonal latitudinal migration the norm, latitudinal migration

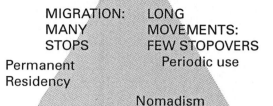

COLD

Seasonal
use

MIGRATION: LONG
MANY   MOVEMENTS:
STOPS   FEW STOPOVERS
Permanent   Periodic use
Residency

Nomadism

WET   DRY
WARMER

Figure 6.2. A simple model of the water regimes and climatic factors that seem to drive various types of movement of wetland birds: regular latitudinal migrants with varying stopover patterns, nondirectional nomadic movements, and permanent residency.

also occurs among waterbirds at the extremes of South America (Weller 1975). The best general correlate with the propensity to migrate long distances in land birds is latitude in the northern hemisphere (e.g., Newton and Dale 1996); temperature extremes would be even more important to wetland species where winter ice precludes the use of most wetlands in much of the northern hemisphere and the extremes of the southern hemisphere. Birds living in extreme cold with seasonal feeding are literally frozen out and must move to more moderated areas, inducing seasonal migration (Fig. 6.2). However, these birds often leave before food resources are depleted, as in shorebirds in the Arctic (Schneider and Harrington 1981) either because of variation at those sites or perhaps related to food availability along migration routes. Climate influences such as temperature are especially important in the timing of water-bird migration as they affect arrival conditions at a breeding or wintering site. In spite of testing over a geological time-frame for some species of migrants, catastrophic events still occur, such as mortality during migration (Fredrickson 1969) and breeding delays caused by late freezes (Hochbaum 1944, Low 1945).

Patterns of movement for a given species also may be geared to reducing competition with breeding birds, as suggested both for shorebirds (Pitelka 1959) and ducks (Salomonsen 1968). Wetland birds living in moderate temperature regimes where water supply provides resources all year would not be forced to migrate and they may maintain year-round territories. However, all the degrees between extreme movement patterns exist, and we recognize that birds leave areas where they could winter (i.e., where temperatures are not so extreme that they could not survive) and go to areas which are farther away than we think necessary. Recently, the recognition that energy demands regulate many such actions suggests that flight may be less costly than constant search effort where foods are reduced in winter; obviously, the unpredictability of resources would have a strong influence on success and survival of nonmigrants. Birds do tap major weather fronts to exploit favorable winds for

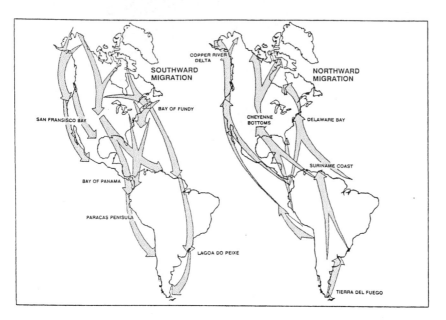

Figure 6.3. Schematic diagram of spring and fall migration of some global migrant shorebirds in the western hemisphere (Helmers 1992, with permission by Wetlands for Americas).

long-distance travel (Bellrose 1957), which can significantly reduce energy demands (Butler *et al.* 1997). In areas where drought (regardless of season) rather than cold makes the wetland unsuitable, movement is essential for water and food, and nomadism would be expected.

Variability in extremes of distance and timing are no doubt the product of long-term trial and error; what works presumably remains as the pattern whether by genetic encoding or by tradition (Hochbaum 1955). The pathways of migration have been studied around the world and often are fairly well-defined routes, especially in the highly social migrations of waterbirds like waterfowl, cranes, and some shorebirds. In North America, major flyways (Lincoln 1935) tend to follow general geomorphic features of the continents such as river courses and coastlines; they also may avoid some major plant communities such as forests that may be inhospitable. However, smaller and better defined routes termed corridors (Bellrose 1968) have been identified and their populations tracked over time; these will be of major concern when we address wetland resources and management strategies below. Several studies of different bird groups have shown the importance of migration stop-over sites, and the predictability of resources common to tidal estuarine systems lures enormous numbers to the same stopovers annually. For example, Hicklin (1987) estimated that up to 1.4 million shorebirds may use the Bay of Fundy in their autumn migration. Many shorebirds are worldwide in their movements, and their general patterns are well documented (Fig. 6.3) (Helmers 1992). Moreover, fidelity of even long-distance migrants to such areas is remarkable; some color-banded Semipalmated Plovers returned to an east coast site for 8 continuous years (Smith and Houghton 1984).

Bird mobility and wetland predictability

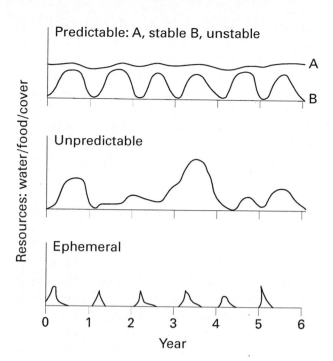

Figure 6.4. Resource variability and predictability that will potentially influence use of and return to wetlands (Weller 1988, with permission).

6.2 Precipitation and wetland influences

Regional precipitation and temperature patterns also may determine the timing of wetland water conditions, especially prominent in shallow waters of temporary or ephemeral pools, spring-fed meadows, and floodplains. Water is the chief driver of all resources in wetland systems because it dictates the timing of seasonal plant growth and animal succession. Where water is predictable, there should be greater assurance of reliability or dependability of food or other resources. Therefore, understanding patterns of regional wetland variation is essential; however such data have only rarely been linked with use, density, or diversity of birds. Predictability of conditions in two widely separated wetland areas used for breeding or wintering would be especially important. However, predictability does not infer stability of resources and could include some pattern of instability (Fig. 6.4, terms from Southwood 1977). It might also include time-limited resources such as invertebrate pulses induced by spring flooding of shallow wetlands or algal blooms, which are synchronized with nutrient availability in deeper water. Unpredictable or stochastic patterns could result in availability of major resources but at an uncertain time and place. These resources would seem less likely to support a regular migratory population, but opportunistic flocks of bird could efficiently exploit them. In Australia and other large arid regions, waterfowl and other birds are adapted for quick response to availability of wet areas via pattern of movement, flexibility in ability to lay eggs over a long period, and by having a large clutch (Briggs 1992).

Changes in waterbird migration in some species have challenged long-held ideas that these movements are all ancient patterns. Cattle Egrets first appeared in Florida in the early 1950s, and by the 1970s had established breeding populations widely from southern Canada to Mexico and Florida. They now even move out of south Texas in winter to Central America (Telfair 1983), where presumably they have more predictable water, temperature, and accompanying food supplies. A similar extension of breeding range of Northern Pintails from drought-impacted or drained prairie wetlands to Alaska has been documented over several time periods, often based on banded (ringed) birds (Derksen and Eldridge 1980). There also have been changes in southward migration routes of Canvasbacks and other diving birds, which have shifted from areas of reduced food supplies in polluted impoundments to richer but more distant areas (Bellrose 1980). Reservoir construction along the Missouri River has resulted in changes in migration routes of Canada Geese, in their choice of wintering areas, and in delayed flights to final wintering sites while they exploit assured resources (Simpson 1988). Other species of long-range migrants have extended their ranges: the Pacific Black Brant, which moves from Alaskan or Siberian nesting areas to coastal estuaries of the west coast of North America, now moves into Mexican estuaries in winter. Its resources are marine, characteristic of fringing wetlands, and its migration is linear along the coast but with long overseas flight as necessary to reach its resources. Stops are relatively few but probably fairly long, like those of some shorebirds, making the protection of these sites crucial for the survival of the species. Most studies of west coast estuaries suggest that the extension to Mexico has resulted from human disturbance in estuarine feeding areas along the west coast (Henry 1980) and reduced availability of eelgrass caused by pollution and human-related disturbance (Wilson and Atkinson 1995). More recently still, the species seems also to be using estuaries in Japan (Derksen *et al.* 1996).

6.3 Influences of wetland type and water regimes

Based on characteristic long-term averages of water regimes shown by various wetland types (see Fig. 2.1), coastal waters and associated wetlands probably are among the more stable and predictable areas, with daily tidal regimes being the most important influence. There are long-term changes in tides produced by lunar and solar patterns, but within a reasonable framework so birds could predict water and common food availability in most years. It is not surprising that many interior as well as coastal breeders move directly to large coastal estuaries to feed on these rich saline resources. Many of these birds are long-distance migrants: Arctic Terns, Red Knots, Black-bellied or Grey Plovers, and many sandpipers. Patterns of migration for these long-range migrants in the

Bird mobility and wetland predictability

western hemisphere have been generalized in Fig. 6.3 (Helmers 1992). Red Knots and many other long-distance migrants cross open ocean in high-altitude flights covering thousands of miles at an estimated 40 to 50 m.p.h (Harrington 1996, Zimmerman 1990) between few but often long feeding stops, where they regain lost weight and then move on. Although Arctic Terns may be seen feeding near coastlines or in pack ice where food resources are reasonably constant, they do make long overwater flights at good flight speeds, where they have little opportunity or inclination to stop (Alerstam 1985). Movements of many such birds have been traced by radar scanning and now by satellite receivers of radio-tagged birds, which has resulted in identification of unknown and isolated resting and wintering areas for species like Stellar's Eiders, which breed in freshwater ponds but winter at sea (Peterson, Douglas and Mulcahy 1995).

Large river systems typically have water in their channels much of the year, but many have a seasonal flooding period and, therefore, great variability in food resources. Birds clearly tap these opportunistically (Douthwaite 1977, Frith 1967), and it is not surprising that major waterbird migration routes follow valleys of large river systems for both daytime and nocturnal migration. The mass migration in the Mississippi River Valley has been recognized as one of the great avian spectacles because of the waterbird migration (Bellrose 1957), which in turn reflects the assurance of resources along its route. Whereas many species follow these linear wetlands, some stop for lengthy periods and exploit their associated resources: Sandhill and Whooping Cranes, White-fronted Geese, and some ducks and shorebirds. The big bend of the Platte River in Nebraska is a prime example, where tens of thousands of Sandhill Cranes, and even more geese and Northern Pintails, gather in spring migration (Shoemaker 1989).

Large lakes or inland seas rank high in stability but their value as habitat for migrants depends on their geographic location and the adaptability of the taxa to deeper and more truly aquatic habitats. Loons, grebes, cormorants, and sea-ducks use these as they would nearshore ocean waters. Many constructed reservoirs function in this way and have become important stopover and/or wintering areas for geese and diving ducks. Other human-induced, deep water bodies like the Salton Sea of Southern California that are located in warmer climes have become major wintering areas for grebes, pelicans, and North American Ruddy Ducks. Constructed reservoirs have shown how Canada Geese and Mallards now move no further south than necessary to where they can winter in open water with rich foods in adjacent agricultural uplands; and diving ducks may exploit benthic invertebrates in these deep and often warmer waters.

Interior wetlands of the Prairie Pothole Region and Great Plains of North America, as well as those of similar landforms in Argentina, demonstrate diverse wetland sizes in which water levels are influenced by seasonal rainfall. These wetlands probably achieve great productivity because of the shallow and dynamic waters, as drying and aeration result in nutrient cycling and natural fertilization. Dependent upon the local climatic regime, many wetlands tend to be wet in spring and early summer and dry later; thus becoming less suitable for waterbirds in autumn or for overwintering; even where temperature regimes would permit use year-round. Movements of birds like ducks, geese, rails, and shorebirds utilize many of these wet areas and often favor small areas in spring and the larger ones that still hold water in the autumn. The regularity of these water regimes probably has been an important influence on strategies of migration, as suggested by shorebirds that stop at the same wetlands but come various distances from breeding and wintering areas (Skagen and Knopf 1993). Not only do some northerly shorebirds use these arid-land wetlands during migration, but residents often are birds like avocets, Willets, and Black-necked Stilts, which can tolerate and even favor shallow saline (i.e. akaline or soda) wetlands. Other migrants follow a "loop" strategy of migration that exploits the resources of the Great Plains wetlands in spring (White-rumped Sandpiper, Hudsonian Godwit, and Lesser Golden Plover) but not in the autumn, when instead they move from breeding areas to the northeast coast where water and food may be more reliable than in the dry plains (Zimmerman 1990), and from where they leave on long overseas flights to South America. A few species follow a reverse pattern (Western Sandpiper), presumably because resources for spring migration are better on the western coast (Zimmerman 1990).

Many shorebirds and ducks make short flights between residual wetlands of extensive complexes, these areas becoming major resting and feeding areas at times (Bellrose and Crompton 1970, Skagen and Knopf 1993, 1994). Presently, there is great concern about human-induced modification of the shallow but interior wetlands that once formed major stopovers for migrant shorebirds and now are drying as a result of alternative water uses (Zimmerman 1990) or drainage (Krapu 1996, Skagen and Knopf 1993). Obviously, society must learn that a wetland which is used only a few weeks or months a year is an essential element of the life cycle of millions of birds treking thousands of miles in a constant search for resources.

It is more than chance that waterbird wintering areas tend to have seasonal rainfall regimes that provide water resources, and that many are near coasts or large interior water bodies that ensure still more dependable resources. The resource tracking pattern of interior breeding waterbirds also involves movements of significant distance and direction as needed. For example, waterfowl from the Prairie Pothole Region or large and saline Great Basin marshes move to winter in coastal freshwater ponds or saline bays along the Gulf of Mexico (Fig. 6.1). Blue-winged Teal move into Central America or northern South America (Botero and Rusch 1988, Saunders and Saunders 1981, Weller 1975), but despite their powerful flight capabilities, few cross the Tropic of Cancer. I suspect that many of the wetlands beyond this latitude are less predictable in their suitability because of the reverse seasonal patterns, which influence whether they can serve as migration stops and wintering areas.

6.4 Nomadism

Birds have been termed nomadic when they are geographically and temporally unpredictable; these birds seek out suitable conditions for nesting, feeding, and other functions (Briggs 1992). Such shifts of populations and assemblages are renowned in Australia (Frith 1967) but also occur in Africa (Douthwaite 1980, Milstein 1984) and India (Ali 1979, Breeden and Breeden 1982). In some cases, drying conditions induce birds to move to new wetlands within the region that are more suitable for nesting. In larger areas and where temperature is nonlimiting, some species seem merely to wait until rainfall results in suitable conditions. Droughts are common to the southern USA where reduced late winter and spring rainfall causes potential breeders such as coots, grebes, and ducks to move out of drying wetlands because they cannot swim and find food and security. Nomadic shifts of breeding ducks and presumably other wetland birds also are common in the western basin and northern USA glacial potholes, with populations moving long distances but often after breeding when both young and adults may capitalize on foods elsewhere. Based on my personal experience in large emergent marshes in Argentina, and discussion with residents and other biologists who have worked there, this type of nomadic response is common. Both in northern Argentina and in the southern USA, movements and breeding of the Masked Duck reflect a species that is unpredictable in nesting location because its habitat requirements are precise. In some way, it seems to locate suitable newly flooded and densely vegetated wetlands or ricefields. When conditions are suitable, the bird may appear, as it has several times recently in southern Texas (Blackenship and Anderson 1993).

The term nomadism has also been applied to regional shifts in breeding colonies of White Ibis, which like Wood Storks have moved out of Florida and northward into Georgia and the Carolinas (Frederick *et al.* 1996). Concurrently, heron and egret populations have decreased in Florida and Texas but increased in Louisiana, possibly because of wetland resources produced by the predictability of extensive crayfish aquaculture (Fleury and Sherry 1995). In such cases, one can only speculate whether this represents growth of newly established populations with concurrent decline of old colonies or actual shifts in populations. Sedge Wrens seemingly have small territories and meet their nesting needs in very small areas, but they must be very mobile to find these because they use such short-lived habitats produced by flooding and drying. They, therefore, are unpredictable (on the basis of our present understanding) in their presence in a prairie wetland and may be breeding in June in one area and in August in another (Bedell 1996). They seem more nomadic within their breeding range but are still long-distance migrants. However, this may well be more the norm than the exception for many inconspicuous birds of shallow and temporary wetlands.

Drought during late summer also can attract post-breeding waterbirds such as Wood Storks, ibises, and egrets to move into drying wetlands to take vulnerable fish and amphibians essential to their breeding cycle. Winter droughts in the same areas mean poor stopovers for migrants and poor wintering areas.

6.5 Local movements

Movements also can be viewed at the scale of a single wetland, as in a swamp where habitat heterogeneity is great or in fringing wetlands along large lakes or seashore. After reaching such an area following long-distance movement, the establishment of small and resource-rich breeding territories, defended or undefended feeding sites (varying by species or group), or the more extensive home range may require only localized movements even when powers of flight are strong (wrens, blackbirds, warblers, and sparrows). It is probably a matter of efficiency but also of body size and resource distribution. Other species, perhaps those more dependent upon water, often build the local habitat unit around a wetland complex that provides various needs but also may act as a backup in event of catastrophic change. In breeding areas, some species seem to have minimal mobility (coots, grebes, loons) so selection of the wetland is crucial; in fact, some water-adapted species like grebes may temporarily lose their powers of flight owing to muscle degeneration. Waterbirds that are shoreline and mudflat-walkers or that wade in shallow water are highly mobile because foods are so closely linked to precise water depths; a centimeter can make major differences in invertebrate distribution.

6.6 Trophic niches, seasonality, and distribution patterns

Many birds move when latitude, temperature, and even water conditions do not demand it, which provides a clue that other factors may influence mobility. Prominent among these is food, especially as tied to life-cycle needs, and fulfilling those needs must be viewed as an optimization process that incorporates various strategies (i.e., feeding adaptations, tactics, mobility, habitat selection, etc.). Among waterfowl, and probably many omnivorous birds, food choice shifts from animal during pre-breeding and annual molt, which supplies needed proteins to plant carbohydrates, which provide energy and fat storage, as needed. Birds use and store fat in different ways, some building all winter (Ring-necked Ducks) and others maintaining minimal body mass until nearer migration (Lesser Snow Geese). But most important here is that these shifts in food may be influenced by abundance and availability in different wetlands – a difficult and complex thing to measure.

Figure 6.5 is a generalization of the patterns of food shifts for waterfowl (Weller 1975), which has been further documented by various studies (see

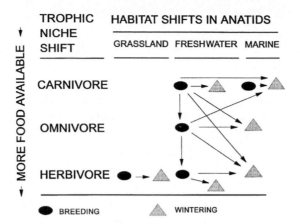

Figure 6.5. Habitat and niche shifts characteristic of North American waterfowl in their typical annual migrations from breeding to wintering areas (after Weller 1975).

Krapu and Reineke 1992). Examples of the three trophic niches may help us to examine similar relationships in other waterbirds. Specialized, year-long carnivores (typically invertivores or piscivores) are species like eiders and scoters, which eat mussels and clams, and mergansers, which eat fish or invertebrates; plant food seems rare. These birds may shift geographic location and habitat but seem to have minimal major changes in foods. Herbivores are fairly common among waterfowl, and their digestive system seemingly allows little use of animal matter. Black Brant feed on foliage of either seagrasses or sedges and grasses at breeding sites and almost exclusively on seagrasses during migration and in winter. Interior freshwater herbivores like geese feed on grasses and sedges and are prominent where these foods are common, even in winter. Southern Screamers seem to feed on floating aquatics during nesting and on upland grasses during nonbreeding periods (Weller 1967).

Omnivory is more common among waterfowl presumably because it is efficient in fulfilling nutrient and energy needs, and in capitalizing on the most abundant food at the most convenient time of year (Fig.e 6.4). Redhead Ducks were cited above as birds of moderate to larger wetlands and more open waters. The diminutive dabbling teal (such as the Green-winged Teal of North America, the Common Teal of Eurasia, and the Speckled Teal of South America) generally favor smaller wetlands for breeding, or the muddy edges of large ones. Teal are small bodied and have tiny lamellae along the edges of their bills that allow them to sort out the tiniest of seeds and invertebrates, making them suitable to use small wetlands of diverse types over a wide range of vegetation and salinity. Their upper mandible is equipped with a nail on the tip and apparently works on large food items too, as they feed on snails and clams and have been observed floating down the Columbia River on dead salmon – eating all the way!

In other orders of waterbirds (Table 3.1), similar ranges of food use and adaptation occur. First, many are carnivores and lack dramatic seasonal food change (e.g., grebes, loons, cormorants, pelicans, herons, shorebirds, flycatchers, etc.). Some change in food size and taxon probably occurs during feeding young, and adults are more flexible in feeding sites because they are less restricted to the nesting area. Their movements naturally are geared to the availability of open (i.e., unfrozen) water of suitable depths for capturing aquatic prey.

As noted among the ducks, many if not most wetland birds seem to be omnivores, and shifts in food are complicated by sites and wetlands types (e.g., coots, rails, warblers, sparrows, blackbirds, etc.). Herbivores seem less common, presumably because use of pure plant foods requires considerable specialization of the digestive system. Perhaps the most specialized is a marginal wetland bird, the Hoatzin, which eats coarse leaves of tropical trees, but its mobility seems to be limited. Some flamingos feed on algae rather than macroinvertebrates, which may explain some of their massive if not long-distance movements to track suitable water conditions. Cranes demonstrate considerable variation in movements based on season and functional needs: many use marshes during nesting, wet meadows for tubers during migration, and upland areas in nonbreeding periods.

Although mobility is common and expected, it differs among groups and much is still unknown. Patterns are especially difficult to measure and interpret when birds are adaptable and opportunistic. It is important, therefore, that interesting observations of even single incidents or suspected relationships be published so that they can be synthesized for pattern identification. Mobility presumably has developed for many reasons, only some of which have been addressed here, and these examples from wetland birds may provide clues to causes of mobility in other birds where the driving forces may be more obscure.

References

Alerstam, T. (1985). Strategies of migratory flight, illustrated by Arctic and Common Terns, *Sterna paradisea* and *Sterna hirundo*. In *Contribution in Marine Science* Vol. 27: *Migration: mechanisms and adaptive significance*, ed. M. A. Rankin, pp. 580–603. Austin, TX: University of Texas .

Ali, S. (1979). *A book of Indian birds*. Bombay: Bombay Natural History Society.

Bedell, P. A. (1996). Evidence of dual breeding ranges for the Sedge Wren in the Central Great Plains. *Wilson Bulletin* 108, 115–17.

Bellrose, F. C. (1957). A spectacular waterfowl migration through central North America. *Illinois Natural History Survey Biological Notes* 36, 1–24.

Bellrose, F. C. (1968). Waterfowl migration corridors. *Illinois Natural History Survey Biological Notes* 61, 1–24.

Bellrose, F. C. (1980). *Ducks, geese and swans of North America*, 3rd edn. Washington, DC: Wildlife Management Institute.

Bellrose, F. C. and Crompton, R. D. (1970). Migration behavior of mallards and black ducks as determined from banding. *Illinois Natural History Survey Bulletin* 30, 167–234.

Blakenship, T. L. and Anderson, J. T. (1993). A large concentration of Masked Ducks

(*Oxyura dominica*) on Welder Wildlife Refuge, San Patricio County, Texas. *Bulletin of the Texas Ornithological Society* **26**, 19–21.

Botero, J. E. and Rusch, D. H. (1988). Recoveries of North American waterfowl in the Neotropics. In *Waterfowl in winter*, ed. M. W. Weller, pp. 469–82. Minneapolis, MN: University of Minnesota Press.

Breeden, S. and Breeden, B. (1982). The drought of 1979–1980 at the Keoladeo Ghana Sanctuary, Bharatpur, Rajasthan. *Journal of the Bombay Natural History Society* **79**, 1–37.

Briggs, S. V. (1992). Movement patterns and breeding adaptations of arid zone ducks. *Corella* **16**, 15–22.

Butler, R. W., Williams, T. D., Warnock, N., and Bishop, M. A. (1997). Wind assistance: a requirement for migration of shorebirds? *Auk* **114**, 456–66.

Derksen, D. V. and Eldridge, W. D. (1980). Drought-displacement of pintails to the Arctic Coastal Plain, Alaska. *The Journal of Wildlife Management* **44**, 224–29.

Derksen, D. V., Bolinger, K. S., Ward, D. H., Sedinger, J. S., and Miyabayashi, Y. (1996). Black Brant from Alaska staging and wintering in Japan. *Condor* **98**, 653–7.

Douthwaite, R. J. (1977). Filter-feeding ducks of the Kafue Flats, Zambia, 1971–1973. *Ibis* **119**, 44–66.

Douthwaite, R. J. (1980). Seasonal changes in the food supply, numbers and male plumages of pigmy geese on the Thamalakane river in northern Botswana. *Wildfowl* **31**, 94–8.

Fleury, E. and Sherry, T. W. (1995). Long-term population trends of colonial wading birds in the southern United States: the impact of crayfish aquaculture on Louisiana populations. *Auk* **112**, 758–61.

Frederick, P. C., Bilstein, K. L., Fleury, B., and Ogden, J. (1996). Conservation of large, nomadic populations of white ibises (*Eudocimus albus*) in the United States. *Conservation Biology* **10**, 203–16.

Fredrickson, L. H. (1969). Mortality of coots during severe spring weather. *Wilson Bulletin* **81**, 450–3.

Frith, H. J. (1967). *Waterfowl in Australia*. Honolulu, HI: East-West Press.

Gauthreaux, S. A. (1982). The ecology and evolution of avian migration systems. In *Avian Biology*, Vol. VI, eds. D. S. Farner, J. R. King, & K. C. Parkes, pp. 93–168. San Diego, CA: Academic Press.

Harrington, B. (1996). *The flight of the red knot*. New York, NY: W. W. Norton.

Helmers, D. L. (1992). *Shorebird Management Manual*. Publication No. 3. Manomet, MA: Wetlands for the Americas.

Henry, W. G. (1980). *Populations and behavior of Black Brant at Humboldt Bay, California*. MSc thesis. Arcata, CA: Humboldt State University.

Hicklin, P. W. (1987). The migration of shorebirds in the Bay of Fundy. *Wilson Bulletin* **94**, 540–70.

Hochbaum, H. A. (1944). *The canvasback on a prairie marsh*. Washington, DC: American Wildlife Institute.

Hochbaum, H. A. (1955). *Travels and traditions of waterfowl*. Minneapolis, MN: University of Minnesota Press.

Johnston, O. W., Warnock, N., Bishop, M. K., Bennett, A. J., Johnson, P. M., and Kienholz, R. J. (1997). Migration by radio- tagged Pacific Golden-Plovers from Hawaii to Alaska, and their subsequent survival. *Auk* **114**, 521–4.

Krapu, G. L. (1996). Effects of a legal drain clean-out on wetlands and waterbirds: a recent case history. *Wetlands* 16, 150–62.

Krapu, G. L. & Reinecke, K. J. (1992). Foraging ecology and nutrition. In *Ecology and management of breeding waterfowl*, eds. B. D. J. Batt, A. D. Afton, M. G. Anderson, C. D. Ankney, D. H. Johnson, J. A. Kadlec, and G. L. Krapu, pp. 1–29. Minneapolis, MN: University of Minnesota Press.

Lincoln, F. C. (1935). *The waterfowl flyways of North America*. US Department of Agriculture Circular No. 342, pp. 1–12. Washington, DC. US Department of Agriculture.

Low, J. B. (1945). The ecology and management of the redhead, *Nyroca americana*, Iowa. *Ecological Monographs* 15, 35–69.

Milstein, P. le S. (1984). A waterfowl survey in southern Mozambique, with conservation implications. *Proceedings of the Pan-African Ornithological Congress* 5, 639–64.

Newton, I. and Dale, L. (1996). Bird migration at different latitudes in Eastern North America. *Auk* 113, 626–35.

Patterson, J. H. (1976). The role of environmental heterogeneity in the regulation of duck populations. *Journal of Wildlife Management* 40, 22–32.

Petersen, M. R., Douglas, D. C. and Mulcahy, D. M. (1995). Use of implanted satellite transmitters to locate spectacled eiders at-sea. *Condor* 97, 276–8.

Pitelka, F. A. (1959). Numbers, breeding schedule, and territoriality in pectoral sandpipers of northern Alaska. *Condor* 62, 233–64.

Salomonsen, F. (1968). The moult migration. *Wildfowl* 19, 5–24.

Saunders, G. B. and Saunders, D. C. (1981). *Waterfowl and their wintering grounds in Mexico, 1937–1964*. Research Publication No. 138. Washington DC: U S Fish & Wildlife Service Wildlife.

Schneider, D. C. and Harrington, B. A. (1981). Timing of shorebird migration in relation to prey depletion. *Auk* 98, 801–11.

Simpson, S. G. (1988). Use of the Missouri River in South Dakota by Canada Geese in fall and winter, 1953–1984. In *Waterfowl in winter*, ed. M. W. Weller, pp. 529–40. Minneapolis, MN: University of Minnesota Press.

Shoemaker, T. G. (1989). Wildlife and water projects of the Platte River. In *Audubon Wildlife Report, 1988/1989*, pp. 285–334. New York: National Audubon Society.

Skagen, S. K. and Knopf, F. L. (1993). Towards conservation of midcontinental shorebird migrations. *Conservation Biology* 7, 533–41.

Skagen, S. K. and Knopf, F. L. (1994). Residency patterns of migrating sandpipers at a midcontinental stopover. *Condor* 96, 949–58.

Smith, P. W. and Houghton, N. T. (1984). Fidelity of semipalmated plovers to a migration stopover area. *Journal of Field Ornithology* 44, 247–8.

Southwood, T. R. E. (1977). Habitat, the template for ecological strategies? *Journal of Ecology* 46, 337–65.

Telfair II, R. C. (1983). *The cattle egret: a Texas focus and world view*. College Station, TX: Texas A&M University Press.

Weller, M. W. (1967). Notes on some marsh birds of Cape San Antonio, Argentina. *Ibis* 109, 391–411.

Weller, M. W. (1975). Migratory waterfowl: a hemispheric perspective. In *Symposium* on wildlife and its environment in the America's. *Publiciones Biologicas Instituto de Investigaciones Cienificas, Universitad Autonoma de Nuevo Leon* 1, 89–130.

Weller, M. W. (1988). The influence of hydrologic maxima and minima on wildlife habitat and production values of wetlands. In *Proceedings of the National Symposium on Wetland Hydrology*, eds. J. A. Kusler and G. Brooks, pp. 55–60. Berne, NY: Association of State Wetland Managers.

Wilson, U. W. & Atkinson, J. B. (1995). Black Brant winter and spring-staging use at two Washington coastal area in relation to eelgrass abundance. *Condor* **97**, 91–8.

Zimmerman, J. L. (1990). *Cheyenne Bottoms, wetland in jeopardy*. Lawrence, KS: University Press of Kansas.

Further reading

Alerstam, T. (1990). *Avian migration*. Cambridge, UK: Cambridge University Press.

Cooke, W. W. (1906). *Distribution and migration of North American ducks, geese and swans*. Biological Survey Bulletin No. 26, pp. 1–90. Washington, DC: US Department of Agriculture.

Douthwaite, R. J. (1978). Geese and red-knobbed coot on the Kafue Flats in Zambia, 1970–1974. East African Wildlife Journal 16, 29–47.

Johnston, D. W. and McFarlane, R. W. (1967). Migration and bioenergetics of flights of the Pacific Golden Plover. Condor **69**, 156–68.

National Audubon Society (1997). The ninety-fifth Christmas bird count. National Audubon Society Field Notes **51**, 135–710.

Skagen, S. K. (1997). Stopover ecology of transitory populations: the case of migrant shorebirds. *Ecological Studies* **125**, 244–69.

Weller, M. W. (1968). Some notes on Argentine anatids. *Wilson Bulletin* 80, 189–212.

7

Other behavioral and physical influences on wetland living

7.1 Territories and home ranges

At some time in the life cycle, most birds seem to space themselves by means of habitat selection and behavior. Within the home range, depending on time of year and functional need, many species of birds have a well-defined breeding or feeding territory that is defended by singing, calling, and/or aggressive chases; some waterbirds also defend stopover sites during migration and winter territories. The origins and advantages of such behavior has been discussed by many workers, and include resource protection, foraging efficiency, predator defense, population regulation, competitive advantage, and energy efficiency (Anderson and Titman 1992, Ashmole 1971, Orians 1971). Many waterbirds are highly social in breeding and feeding, and although territories may be clear-cut and violently defended, there is a wide range of territory size and behavior among wetland birds. While there is a positive relationship between body size and territory size in birds (Schoener 1968), it also seems to be dependent upon nesting and feeding strategies. Defense of territories that incorporate most life functions, such as mating, nesting, and feeding (see classification in Pettingill 1970), is most common among species that for reasons of food supply, water access, or nest cover restrict themselves to small but diverse patches and move little during the breeding period, e.g., some grebes and loons (Bergman and Derksen 1977), or coots (Gullion 1953). Species that nest colonially, such as egrets, herons, ibises, cormorants and pelicans, commonly defend only a small area of the tree, shrub, or marsh vegetation or substrate that is a mating site as well as nest site. However, they are attracted to and rarely nest outside colonies, and they obtain foods at some distance from the site, where they may be highly social in feeding. A few shorebirds have mate-only territories or leks (Oring 1982).

Grebes are good examples of aggressive birds that demonstrate a wide range of behavior by species. Some are solitary (Pied-billed Grebe, Great-crested Grebe), and some are colonial (Eared and Western Grebes). There also is a great range of agonistic behavior among ducks, with dabbling ducks typically showing more aggressive behavior than bay ducks (pochards) or seaducks. Blue-winged Teal and Shoveler are particularly aggressive on relatively small territories where they typically feed on invertebrates (McKinney 1965). Several species of flightless duck, also strongly carnivorous, tend to use sections of seashore and probably are permanent residents, for example steamer-ducks of southern South America, and Auckland Island Flightless Teal (Weller 1980). Similar defense patterns and territory size relationships are evident in species that use streams and where foods are dispersed by virtue of the physical characteristics of the sites, for example dippers (Tyler and Ormerod 1994), Blue Ducks (Eldridge 1986), African Black Ducks (Ball *et al.* 1978), and Torrent Ducks. Common Eiders are known for their large colonies along seashores whereas Stellar's, King, and Spectacled Eiders normally nest alone or in loose nesting clusters (Anderson and Titman 1992, Bellrose 1980, Delacour 1954–64). There is an advantage in not having a small and defined territory for parasitic birds like Black-headed Ducks (Weller 1968).

Territorial behavior also occurs on feeding areas for migration and wintering. A number of groups, like shorebirds that feed on dispersed foods such as invertebrates buried in mud or sand, may defend an area year-round (Myers, Connors and Pitelka 1979). Rock Pipits, which feed in the intertidal zone, defend a winter terrritory centered about food resources (Gibb 1956). Fish-eating birds like loons are known to "roost" together at night but disperse and defend feeding areas during the day (McIntyre 1978). Coots and some water-fowl that are permanent residents seem to defend their territories year-round (Gullion 1953, Weller 1980).

Male defense of females or parental defense of young seem more common than site defense in some more social wetland species like geese, but many behaviorists do not regard this as true territoriality because it does not involve a geographic area. Defense of family members or feeding sites in species with long-lasting parental ties also is common among geese and cranes on migration stops and wintering areas. Some wetland birds seem to defend feeding areas more aggressively than breeding areas, but this may be a result of their more conspicuous social behavior during nonbreeding (e.g., shorebirds). In contrast, ducks are known for their mixed-flock feeding on invertebrates and seeds in water areas where resources are evidently very abundant, especially during autumn and winter, and fighting is not conspicuous. Presumably, aggressive behavior is reduced by niche segregation of species involving use of food of varying sizes or different prey, which in ducks seems related to lamellae spacing and body size (Nudds and Bowlby 1984). Mixed flocks of herons, egrets, and ibises show mainly feeding-area defense, but often these birds are preying on superabundant and vulnerable prey of different sizes.

7.2 Breeding chronology, phenology, and ranges

Various bird species nest at different times of the year as a result of patterns of environmental seasonality, laying patterns, incubation period, and growth rates of young. But presumably all seasons are influenced by adaptations to available resources in chosen habitats, and how rapidly they are depleted (Wittenberger and Hunt 1985). In an assemblage of birds, there also is the probability of competing with a similar species using similar resources (Cody 1981); therefore, temporal segregation such as timing of nesting and rearing young would be an advantage. Another way in which direct resource competition is avoided is segregation according to microhabitats (e.g., pipits in Europe (Svardson 1949) and Red-winged and Yellow-headed Blackbirds in North American (Weller and Spatcher 1965)). The origins and ultimate or evolutionary implications of such arrangements are many, but we are concerned here with how they may be tied to habitat resources. If one examines variation in breeding times of the same species at various places, considerable variation is evident, although establishing why is more difficult. Temperature and other seasonal influences are reflected in year-to-year variation in breeding phenology (Rohwer 1992) but regional climatic differences probably produce different mean nest-initiation dates in each region, regardless of habitat. However, it is common for waterbird migration and/or breeding to be influenced by melting of ice at extreme latitudes, rain in fields that provide food resources (Hochbaum 1944), flooding of marsh vegetation into better feeding areas (Weller 1979), or tide levels that influence breeding cycles of invertebrates, which will result in available food for migrants (e.g., the synchrony of Red Knot migration in relation to egg-laying by Horseshoe Crabs (Harrington 1996)). Some of these events are annual but still vary with the weather, and others are long-term and predictable events (tide cycles) that induce lasting behavioral adaptations. Response of terns to food supplies near oceanic and equatorial islands suggests that occurrence and breeding may be linked to food supplies and probably body condition – without regard to season (Schreiber and Ashmole 1970).

Two examples of quite different species may provide insights about the origins of timing of breeding that relate to habitat choice and specific breeding range. Mottled Ducks are year-round residents of marshes of the Gulf of Mexico near the coast and presumably originated from a Mallard-like ancestor. It seems obvious that this bird has survived in areas where wetlands often dry out in early summer by living where they can shift habitats from freshwater wetlands (best for rearing young because of less salt stress) to coastal marshes that are always wet and rich in foods. Even these coastal marshes are not stable however, because tide cycles do vary and rainfall modifies freshwater inflow so drought-stress periods can occur (Stutzenbaker 1988). Coincidentally, Mallards winter in part of the range of Mottled Ducks, but their breeding cycles do not coincide. Mallards may mate in the fall and early winter but remain in flocks and migrate north to establish territories and breed. Mottled Ducks may flock and migrate but most remain on home ranges as pairs all year, select small pools in large marshes or small wetlands for isolation and feeding, and breed in late winter or very early spring (Stutzenbaker 1988). This results in hatch and growth of young before many of the freshwater ponds dry. Therefore, the two species can coexist for part of the year, using slightly different microhabitats, and show little or no overt aggression where they overlap. An interesting but dangerous experiment was the release of large numbers of hand-reared but wild-stock Mallards in the range of the Mottled Duck in South Texas (Kiel 1970). After many years, Mallards have failed to survive in free-living, viable populations, suggesting that they probably do not do the right thing at the right time in this semiarid area. Because the typical water regimes in northern wetlands involves spring breeding by Mallards, those trying to breed in South Texas in spring and early summer often would meet with drying freshwater ponds, and young would be placed in jeopardy seeking new foraging areas. Mallards do use estuarine areas in some places, but most populations seem to avoid saltwater when possible. Perhaps young Mottled Ducks may be better adapted to saltwater than Mallards, but they do seem to favor freshwater habitats until they can better tolerate salt after the first few weeks of life.

The Wood Stork, a species often thought of in terms of lush and wet tropical wetlands, actually is linked to declining water regimes. It feeds its young on fish and invertebrates, often obtained at long distances from the nest. Its breeding time and location are dictated by water regimes in wetlands, regardless of rainfall. Drying water areas make prey vulnerable to capture, whereas flooded areas make prey less available (Kahl 1964). Nest success is highly variable because of these wetland dynamics, and shifts in populations have been attributed to excessive water in feeding areas. Like many colonial waders, they commonly engage in post-breeding movements of hundreds of miles or more northward to suitable feeding areas, exploiting the foods of drying wetlands.

7.3 Geographical range

Although local and regional habitat dictates presence or absence of a bird species, there are evolutionary (historical) origins and other environmental aspects of regions that influence the distribution of major taxa of birds, which are treated in detail elsewhere (e.g., Vuilleumier 1975). Obviously, there are relatively few higher taxa (orders) that are worldwide, more families or genera that are found on several continents, but few species that are worldwide among birds. Without attempting a detailed analysis, waterbirds must rank high in average size of the geographic range and in distribution on several continents because they are strong fliers, can use small and isolated habitat islands (wetlands), and often are adaptable to a wide range of foods and breeding sites (Weller 1980). However, owing to speciation during geographical isolation, closely related species of the same genera do cover extensive areas of the world as breeding habitat: pintail and Mallard ducks, Black-headed or Brown-hooded Gulls, and several terns and shorebirds. Numerous shorebirds breed widely in the northern hemisphere and winter widely in the extreme southern latitudes.

Analyses of various taxa and continents from an ecological perspective have been attempted for Anseriformes (Weller 1964, 1975, 1980) and Podicipidiformes (Fjeldsa 1985), and a still wider ecological and taxonomic analysis of waterbirds of South America has revealed some interesting patterns of higher species richness of some groups in the low latitudes (tropical fish-eating birds) and of omnivorous waterfowl at subtropical or temperate latitudes (Reicholf 1975). From a habitat perspective, these patterns can be analyzed in only rather general ways because we have too little information on individual species and because we are unable to assess the historical influence that camouflages present habitat issues. They are of special interest, however, because we can see how similar wetland habitats in different areas are used by different taxonomic groups, such as the seeming ecological equivalents in northern and southern hemispheres: Black Brant and Kelp Goose, and Common Eiders and Flightless Steamer Ducks. Additionally, we need to consider interfamily and interorder competition for resources, which probably occurs in some groups of wetland birds and may influence community composition (Weller 1980).

Geographical patterns of bird distribution in wetlands lead one to speculate on the importance of historical events and dominance of highly competitive bird groups in the origin of current distribution. In North America, blackbirds (Icteridae), and wrens and sparrows dominate the songbird community of emergent marshes, with highly social blackbirds being most abundant and conspicuous. In South America, a greater diversity of blackbirds occurs (but few are colonial), but other species (representing several other families not found in North America) nest over water in emergent vegetation. In Europe, Icterids do not occur and there are at least ten "warblers" (Muscicapidae and Silviidae) bearing such common names as "swamp," "marsh," "reed" etc. because they nest and feed in emergent vegetation. In North America, distribution patterns within wetlands range from that of the near-shore or shallow-marsh species like Sedge Wrens or Swamp Sparrows to that of species that like more open and deeper water like Yellow-headed Blackbirds building nests in robust plants like cattail and bulrushes (Fig. 1.1). Clearly, different groups have exploited these rich wetland habitats all over the world; if one taxon or specialist does not use a habitat, another does.

7.4 Pair bonds and other breeding behavior

Mating is permitted by the establishment of at least a temporary pair relationship and takes many diverse forms (Oring 1982). Clearly, habitat plays a major role in the setting necessary for pair associations in the same place where displays, mating, and identification of the nest site is possible. Courtship displays and copulation in the more aquatic waterbirds tends to take place on water. Duck and grebe courtships are renown for their fascinating but often confusing display signals, interesting vocalizations, and beauty of form and color. Shorebirds use more terrestrial sites, have rapid, moving, and agressive actions, and some have aerial displays like the "bleating" or "winnie" of the Wilson's Snipe (a sound created by the tail feathers) or the aerial breaking sound of American Woodcock (resulting from the special wing feathers). Some shorebirds have elaborate display feathers like the collar of the Ruff. Waders and coast-nesting seabirds commonly call and display from potential nest sites to attract a mate. In such species, copulation occurs on the platform or nest site and is tied to nest construction activities such as the ceremonial displays when nest material is brought to the nest. Males of more terrestrial areas and marsh-edge birds like Red-winged Blackbirds or Swamp Sparrows typically establish territories by use of song perches, but the females check out the vegetation, the general locale, and probably the food supply carefully before establishing a site or pair attachment.

Multiple mating systems increase in wetland birds compared with other terrestrial species. According to Verner and Willson (1966), only 14 of 291 North American passerines are polygynous, and 13 of these are wetland birds. There are numerous cases in other taxonomic groups as well and these have been the focus of major studies on the general aspects of breeding in birds. In most species of polygamous marsh songbirds that have been studied in detail, males defend large territories and attract and mate with several females (e.g., North American Marsh Wrens (Verner and Engelson 1970) and Red-winged and Yellow-headed Blackbirds (Orians 1980)). In such territorial species, those with the highest quality territory (i.e., habitat resources) seem to attract the most females, which has been attributed mainly to food availability. Studies in South and Central America suggest that the lower density of food resources may explain the reduced coloniality of blackbirds there compared with birds nesting in North America (Orians 1980).

Although virtually all types of mating system are known in shorebirds, polyandry, the mating of one female with more than one male, is especially common. It may not be influenced by habitat, but it is at least known mainly from this group; however, strategies vary. In some species of sandpipers and plovers, the male of the pair incubates the first clutch and the female the second, presumably taking advantage of optimal resources in a short time frame. Female Spotted Sandpipers, which feed on small, temperate-zone mudflats, lay a clutch of eggs, which a male then incubates. Later, she may lay other clutches if conditions are suitable, and males remain ready to take over the incubation duties. Presumably, locating males in such dispersed habitat patches is difficult and females defend territories around the males. It is assumed that resources are very important, especially the number of invertebrates, which is essential for egg-laying (Oring and Knudson 1972).

Among jacanas, females defend a territory with several males, which incubate the eggs she lays in nests built by the males (Emlen, Demong and Emlen 1989). In Red and Red-necked Phalaropes, females are more colorful in plumage, have prominent courtship displays, and defend territories. Males incubate the eggs and females devote no time to protection of the highly precocial young.

7.5 Nest type and construction

Surprisingly, we probably lack sufficient descriptive material on nest and nest material to compare many species, but information on typical sites is available for the more common birds to indicate general patterns (see references in Chapter 3). Although there seems to be a general relationship between anatomical adaptations for water and the location of the nest and the type of vegetation used, there are many exceptions.

Many ducks, geese, and shorebirds that obtain food for their young from water nest in the uplands, some in very dry places quite some distance from water. After hatching, they move their young to wet areas to feed and often to roost. Some wetland birds, such as pelicans, some gulls, terns, and plovers, do little if any nest building, nesting on sand or gravel substrates. Killdeer sometimes nest on rooftops (an elevated, flat and sometimes-flooded "island") where predator protection is good but where food resources must be quite limiting. The Egyptian Plover or Crocodile-bird (actually a member of the family of pratincoles and coursers rather than a typical plover) is among the more interesting of the shorebirds in nest site and incubation behavior; it reputedly lays its eggs in the sand along the Nile and other North African rivers and covers them to allow the sun to incubate them during the day while the bird takes over at night (Austin 1961). It is well known for its habit of finding foods in the open mouths of sunning crocodiles!

Some birds that nest over the water in emergent vegetation seem to be species that only recently (on a geological time scale) have entered the water environment or that are specialized on this edge because it is such a rich and flexible habitat which can be exploited without much specialization. Even when nests are built over water, such species may carry shoreline plant materials to the site, either by need or heritage, as is true of kingbirds and some blackbirds. Red-winged Blackbirds weave elevated, cup-like nests, often from material at or near the site, but line them with fine grasses, sedges, or even animal hair obtained in the uplands. Yellow-headed Blackbirds (which abandon sites that dry up) use last year's soggy vegetation, which they can weave into a porous basket; however, these birds typically lack the fine land-based materials used by the Red-winged Blackbird (Weller and Spatcher 1965). Some tree-hole nesters like Wood Ducks use tree holes near or in water but are related to the wetland only by the fact that a nest over water can provide protection from ground predators and will reduce the travel time to take their young for foraging. Only down is added to the decomposing tree cavity to enhance an already ideal nesting environment (Breckenridge 1956). Cormorants, herons, egrets, and ibises may build nests in large colonies on elevated sites (islands, shrubs, snags, trees, cliffs), carrying material and fighting over it and stealing it from neighbors.

The more aquatic swans, geese, and ducks that nest over water in emergent plants all use materials at hand (obviously selecting sites with suitable materials) and those nests on land are built of fine grasses and sedges and are down-lined. Those over water may be raised in the vegetation, more by the bending down of supporting plants which criss-cross over the water. As the bird sits on this sparse foundation, it pulls material into the center and molds it into an egg-holding as well as insulating structure. Some build an enclosing canopy over the top by reaching up and pulling down tall emergent vegetation as it grows (Redheads and Ruddy Ducks).

Highly aquatic grebes may build floating nests of wet plant debris derived from the previous year's growth, or sometimes living submergent vegetation. Such nests are quite often in open areas, and coot and grebe nests may suffer high failure rates through large or continuous waves (Boe 1993, Fredrickson 1970).

One of the most unique nests and nest sites among waterbirds is a source of curiosity not only in its origin but also in its functions. Nests of Horned Coots are built of submergent vegetation on slightly submerged mounds of rocks in water about half a meter (2 ft) deep in Andean lakes at elevations of over 3658m (12000 foot) (Ripley 1957). Birds also have been seen carrying large rocks from the shoreline to deposit on 2 ft-high rock islands, which subsequently serve as sites to deposit vegetation and lay eggs. In shallow water, natural elevations of the basin may be used. After nesting, other bird species like grebes, gulls, and ducks may use these constructed sites. The closely related Giant Coot (actually a little smaller than the Horned Coot) is essentially flightless and nests in still higher lakes (up to 6645 m or 21800 ft) (McFarlane 1975). Nests are build of floating masses of aquatic plants, which are protected from drifting by placement on underwater island-like undulations of the basin.

7.6 Nest microhabitat

What influences the location of the nest in relation to habitat features? At the proximate level, feeding success by the potentially nesting female must be of primary concern. Availability of nesting materials, structural support, or cover would be important in selecting the general area and the specific nest site. These are all part of resource quality or habitat suitability. The role of microclimate has not been carefully explored for many wetland birds, but technological advances now allow detailed monitoring of climatic conditions.

The nesting strategy varies with the social structure, as territorial birds protect their nests and often feeding areas. Where polygamy is involved, several nest sites may be protected, but presumably it is the nest builder (whether female or male) that selects the specific nest site within the territory (e.g., in blackbirds, sparrows, and shorebirds). Nesting females with solitary nests and which incubate alone presumably pick and protect their nest sites, often with little involvement of the mates (e.g., ducks). Species with strong pair bonds that share in incubation (Killdeer, stilts, avocets) share in alarm and defense. Our question here is whether these patterns relate to habitats, directly or indirectly, or are merely reflections of a general reproductive strategy common to the ancestral stock and subsequently adapted to wetter habitats. One probably cannot be separated from the other (Southwood 1977), but I suspect the influence of reproductive and foraging strategies blend in some optimized pattern that works for the species or group: feeding strategy seems to dictate territorial and social behavior, with seasonally carnivorous birds like songbirds and shorebirds defending feeding areas that provide for young. But even some herbivores are highly protective of nest sites (e.g., geese). Waterbirds feeding on dispersed food resources generally seem less aggressive. Body size must be a factor in terms of protection from potential predators and hence placement of nests, for a Canada Goose would be less endangered by nesting on the ground than a duck or shorebird. Egg size and size of the young must have a similar influence in relation to the size of the predator, e.g., a small garter snake is serious for a blackbird or sparrow but it would take a crocodile to influence a goose nest site. Many studies of ground-nesting birds have shown a preference for dense over light vegetation cover; the latter may result in greater visibility. One study reported differences among two species of duck in avoiding sun or favoring humidity but found other features of greater importance in dictating nest site (Gloutney and Clark 1997).

7.7 Egg and embryo adaptations

Eggs of birds that nest over water are conditioned to high humidity, and artificial incubation of duck eggs requires either special high-humidity incubators or spraying the eggs with water to prevent drying (addling) of the embryo within the egg. Many grebe nests are a floating mass of dead but wet plant debris or fresh submergent vegetation, and eggs are always wet. Ducks that nest over water and other swimming birds like coots and gallinules probably bring water back to the eggs after each incubation break via wet plumage. However, species like skimmers that nest on dry, tropical islands or shores seem to wet their feet to bring water to the eggs (Turner and Gerhart 1971). Terns and other birds nesting on muskrat lodges presumably have a moisture-retaining site (Nickell 1966).

Egg destruction in nests of other nearby birds is common among Marsh

Wrens as an apparent mechanism to reduce competition for habitat resources. A recent discovery is that eggs of Marsh Wrens are structurally stronger than expected for their size, possibly an adaptation to reduce the potential impact of pecking by other Marsh Wrens (Picman, Pribil and Picman 1996). Presumably, it is a more serious threat to Marsh Wrens than to other species, or one might expect similar egg adaptations among blackbirds and sparrows.

7.8 Precocity and plumage of newly hatched young

In 5 of the 13 orders of wetland birds listed in Table 3.1, there are taxa which have young that are precocial, down-covered, and able to leave the nest within a few days. Of these, the more agile feed themselves and can swim or at least move away from the nest site – often over land (ducks/geese, shorebirds, some rails, cranes). Downy Black-headed Ducks leave the nest of their foster parents within a day or so of hatching (Weller 1968), and many other species of duck-lings are precocious and leave their parents early in life, falling into the higher levels of precocial classifications (see Erlich, Dobkin and Wheye 1988, Skutch 1976). Others such as loons, grebes, rails, and coots tend to remain in the nest longer, move best in water around the nest site, and rely heavily on parental feeding for several days. Gulls and terns stay near the nest longer and are fed directly by their parents for some time. Pelicans, cormorants, herons, kingfishers, and passerines are altricial, being fed in their nests until capable of flying, swimming, or walking away. In most cases, precocial young are also feathered (ptilopedic) and the more altricial species are naked or nearly so (psilopaedic) (Pettingill 1970). Although such patterns do not differ clearly from the range in terrestrial birds, I suspect that there are more precocial species among wetland birds to take advantage of dispersed food resources independently, and because of the need to move if water dries up. Growth rate may be more rapid for the same reason, an especially important asset to avoid freeze-up of young loons and swans reared in the Arctic. Difference in plumage tracts also occur, with juvenile down developing early and feather patches that protect the water-exposed ventral area. Terrestrial groups like quail and pheas-ants develop back and wing feathers more rapidly than those on the ventral area, whereas waterbirds use water rather than flight to protect them from predators.

7.9 Plumages for swimming and diving

Because of the diverse phylogenetic origins of waterbirds, all are not adapted to problems induced by the water environment in the same way. One of these problems relates to keeping dry when swimming and diving. Some deal with this by avoiding swimming (frigatebirds and terns) and others by minimizing time in the water. Anhingas, cormorants, and screamers spend considerable time "drying" their wings in the behavior termed sunning or, more appropriately, wing-spreading (as they will do it in the shade or on cloudy days in breezy sites) (Clarke 1969, Elowsen 1984, Hennemnann 1982, Mahoney 1984). While this drying action undoubtedly lightens a bird for flight, wind-spreading also is done by Flightess Cormorants, suggesting that the water medium is important.

Plumages often differ among well-adapted waterbirds, but little seems to be known about its form and function. Skilled swimmer/divers like loons and grebes have fine and short but dense feathers, which give a patchy-looking plumage that must be specially built for water. Grebes, among the most skilled of divers with body compression and foot design for managing depth and speed of diving, have seemingly wetable plumage (looking ruffled) that they fluff via rise-up body shakes and preen regularly.

Even well-adapted ducks vary in their behavior in relation to water. Dabbling ducks probably spend less time in water that do those of other tribes of waterfowl, and more time preening. Eiders and other sea ducks are seemingly better adapted to long periods in water, aided by down, known for its remarkable air-trapping structure and insulation, but all actively preen even when in water. Bay ducks seem to lie somewhere between, as they spend considerable time in water and may preen on shore, but also preen in water in a belly-up posturing that looks awkward but seems to work. Rails have soft, dull, and airy feathers seemingly poorly adapted to water; however, the plumage is dense in coots, which swim and dive regularly. Cormorants and anhingas spend much of their time in water but have plumages that are quite wetable. Seemingly, it is the air within the feathers that keeps water from penetrating to the skin, and perching out of water and preening are important. Moreover, some workers believe that wing-spreading displays in these species actually are drying actions taken after diving.

7.10 Body size

The significance of body size in birds is a current interest in ecology as it relates to niche segregation (Nudds and Wickett 1994), distribution and rarity of species (Gaston and Blackburn 1996), and population size (Blackburn and Lawton 1994). Its importance here is the need to search for resource-related patterns. Body size of waterbirds varies enormously, from fish eaters like pelicans (up to 15kg) and storks to large herbivores like screamers, swans, and geese (>10kg) to minute rails and passerines (<15g). This range and complexity allows for little analysis except between ecosystems, but patterns within taxonomic groups may be more meaningful. Among waterfowl, herbivores like swans and geese typically exceed the size of the carnivorous sea ducks, which in turn are larger than most omnivorous ducks. There are some general

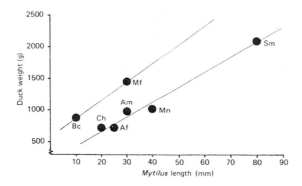

Figure 7.1. Food size of some sea and bay ducks in relation to body size. Bc, Goldeneye; Mf, Velvet Scoter; Ch, Long-tailed Duck; Af, Tufted Duck; Am, Scaup; Mn, Common Scoter; Sm, Common Eider, (Modified from Pehrsson (1976), with permission.)

relationships between body size and prey size in herons (see reviews in Weller 1995) and ducks (Nudds and Bowlby 1984, Pehrsson 1976) (Fig. 7.1), and probably in shorebirds (Holmes and Pitelka 1968) – although opportunism strongly influences this pattern. Although similar relationships are known in land birds, the patterns are often quite dramatic among wetland birds because of the diverse food sizes and types, and resulting specializations of bill, feet, and foraging behavior.

7.11 Sociality: feeding, nesting and resting associations

Many species of birds are social: some only during nonbreeding periods, some all year. In fact, it may be more difficult to identify solitary species of waterbirds, for example numerous raptors, bitterns, usually Limpkins, rails, loons, and certain shorebirds, grebes and waterfowl. But the social ones are many, including most cormorants, pelicans, gulls, terns, grebes, waders, many species of shorebirds, ducks and geese, and New World blackbirds (Lack 1968).

What determines the interspecies associations, groups, or assemblages of waterbirds? Much study has been devoted to intraspecific agonistic behavior in breeding terrestrial birds and seabirds that nest on islands and cliffs (see Lack (1968) for review), for example skimmers (Burger and Gochfield 1990), ducks (McKinney 1992), moorhens (Howard 1940), coots (Gullion 1953), gulls and plovers (Barnard and Thompson 1985), and the heron group (Kushlan 1981, Kushlan, Morales and Frohring 1985). Why do they flock, what resources permit it, and what are its advantages and disadvantages? Is this pattern habitat related or more specifically food related, for predator protection, or for pair relationships? Probably all of the above are factors (Bennetts and Dreitz 1997, Wittenberger and Hunt 1985), and not for all species for the same reason at the same time of year. But here I want to deal more with social flocks of birds as they relate to habitat use and food resources rather than the pair relationships. Some examples are from single species that nest in colonies or feed in groups, but wetlands seem to attract flocks of mixed species because of the concentra-

Behavioural and physical influences

tions of foods under certain conditions that make feeding less competitive, less defensible, and mutually advantageous.

7.11.1 FEEDING

Clearly, freshwater habitats are rich and diverse in food resources and nesting sites. Concentrated food is difficult to protect from congeneric competitors, and such effort probably is not energy efficient except when nesting requires the insuring of a food supply for the young. Moreover, food production in many wetlands seems to be one of seasonal and/or hydrologic pulses, with invertebrates available during mid to late summer and seed levels high during autumn and winter. Vertebrate foods like amphibians and fish often are vulnerable after the production of young and the decline of water levels: this varies seasonally and regionally. At these times of abundant resources, flocks of the same species and multispecies flocks may result.

It is difficult to identify patterns among these flocking species because of regional variability and because of exploitation of opportunities for concentrated resources. Several studies have focused on such patterns by comparing food habits of specific taxa. In Anseriformes, flocking tendencies are highest in those species that are dominantly herbivores like geese, which roost together and feed in large flocks, whereas carnivores are more likely to roost in smaller groups or feed alone (Ydenberg and Prins 1984). Researchers concluded that solitary search for mobile prey and individual competition influence these social patterns. Such patterns vary according to resource need; for example terns feed alone because of food dispersion but are social in nesting or roost areas. There are many alternatives, but feeding tactics and food resources must both play important roles in the evolution of such flocking behavior.

Of the birds listed above, many of the species that lead single life styles tend to be carnivores, feeding by predation on insects, snails, mice, fish, frogs, or birds all year, where individual action is essential and foods may be less concentrated. Many of the other flocking species are of three types: omnivores that switch trophic niches seasonally (e.g., many ducks, blackbirds); herbivores that eat plant foliage and seeds most of the year (geese, sheldgeese, and a few ducks); and carnivores that feed as groups on vulnerable schooling fish, amphibians, or large invertebrates (egrets, herons, cormorants, pelicans, mergansers). But there are many exceptions induced by opportunity. For example, many migrant shorebirds are social in feeding and are carnivorous, feeding mostly on small, abundant, and dense invertebrate prey. Normally solitary Snail Kites, Boat-tailed Grackles, and Limpkins occasionally are seen in company with White Ibis and Great Egrets at low water levels, essentially using them as "beaters" to disturb prey for hunting (Bennetts and Dreitz 1997).

Mixed-species flocks of feeding waterbirds are most common and best known among pelicans, cormorants, ducks, sandpipers, egrets and herons, and gulls. Pelicans and cormorants seem regularly to feed in groups and are joined by several species of gulls. Often they are in masses of mixed species of individuals feeding independently; others seem more organized forming semicircles, rows, and wedges that may drive fish into more vulnerable positions. Presumably, this cooperative effort aids all in locating the schools and perhaps in capture. Sometimes they are linked to porpoise activity, suggesting that all are eating from the same school of fish. Moreover, when fish are large, the leaders seem to be the cormorants while the White, and occasionally Brown, Pelicans are stealing fish from the cormorants.

It is obvious that "mixed" flocks may be a human perception, and that although the bird species are close together they are in small clusters keeping visual and perhaps vocal contact, and often even take flight in groups by species. This is especially evident among sandpipers and other shorebirds feeding side by side. Mixed flocks also occur among aerial predators like swallows, which feed over marshes.

Gulls and terns seem often to be closely associated in feeding, but their feeding tactics differ greatly, and the degree of food overlap probably is minimal except among gulls that hawk for insects. Small terns feed on smaller fish than do large ones, and some feed on surface or emerging invertebrates. Large gulls seem more flexible and scavenge or take prey of all sizes in competition with smaller gulls, but large prey seem most likely to be preferred.

Swan or shoveler and phalarope associations occur commonly, in which case they presumably are competing for the same but very abundant foods only with shoveler. Snowy Egrets and Tricolor Herons have been observed feeding symbiotically with Red-breasted Mergansers or Pied-billed Grebes (Emlen and Ambrose 1970); clearly the diving bird could be taking deeper fish or invertebrates disturbed by the waders searching for surface foods. Relationships between birds and aquatic mammals were mentioned in Chapter 5.

Another interspecies mix involves food stealing (piracy or kleptoparasitism) and is common among waterbirds. Gulls and terns commonly rob food from other species as well as their own, and frigatebirds obtain much of their food that way. A well-studied relationship between Black-headed Gulls stealing from Golden Plovers and Lapwings is common in wet meadows (Barnard and Thompson 1985). Surface-feeding wigeon and Gadwall ducks often harass diving coots as soon as they surface with herbaceous foods and retrieve the dropped items (LeSchack and Hepp 1995).

7.11.2 NESTING

During the breeding period, colonies of some species of gulls, ibises, egrets and herons, and flamingos may contain thousands of nests with spacing dictated by vegetation or other structural support and their territorial pecking distance. However, nesting coloniality does vary regionally, as many South American blackbirds are less social that those in North America (Orians 1980), as is true also of Brown-hooded Gulls of Argentina versus Franklin's Gulls of the northern USA and Canada (Burger 1974). Other factors such as interspecific aggression may influence nest spacing within favored nesting areas; for example, Red-winged Blackbird nests are thought to be clumped as a strategy to prevent egg destruction by Marsh Wrens, which occupy the same habitat (Picman 1980).

Interspecies association of nesting birds has been reported widely where various swimming waterbirds nest preferentially among colonies of gulls and terns in islands of emergent vegetation. Examples are diving ducks (Anderson 1965, Hochbaum 1944), skimmers (Burger and Gochfield 1990), and grebes of several species (Burger 1984, Nuechterlein 1981), which are common in tern or gull colonies. Such associations are thought to reduce aerial nest predation on the duck eggs, but it is difficult to sort out the general attractiveness of such habitat patches to all birds that nest in emergent vegetation over water.

Defense of nests and young often is enhanced among dispersed or colonial birds by virtue of alarm calls and mobbing behavior of birds associated in the same habitat type (Bergman, Swain and Weller 1970). This mobbing behavior is common to many land as well as water birds, but it is well known among wetland birds because of their social nesting behavior. A snake, snapping turtle, or mink can cause great havoc in a marsh but its presence is obvious at great distances through the aerial displays and special alarm calls (Bergman *et al.* 1970). Lapwings and Southern Screamers are kept as pets in places because they are alert and vocal about intruders, exceeding watchdogs as alarm systems. A comparative study of grebes nesting either within or outside gull colonies in Argentina indicated that not only did grebes respond to warning calls of Brown-hooded Gulls but also they had lower predation on adults and eggs and higher nesting success (Burger 1984). If such females return to nesting areas, high success in one year should induce a return to the same site another year.

7.11.3 ROOSTS AND RESTING BEHAVIOR

Roosting is a general term for nocturnal resting of individual or flocks at specific sites (often on water but also in grass, shrubs, or trees), whereas daytime groups are termed resting flocks. Its commonality is that both activities reflect a choice of some form of social grouping of individuals (and sometimes mixed species) that performs a sometimes unclear function (Wittenberger and Hunt 1985). Its relevance here is that such social gatherings require special habitat and certain conditions within that habitat to meet the needs of a species or mixed flock. Some birds that are typically solitary species may roost in groups at night during migration (e.g., kingfishers (Fry, Fry and Harris 1996) and wagtails (Zimmerman, Turner and Pearson 1996), or in open water on wintering areas) (loons) McIntyre 1978)). Roosting sites may involve trees or shrubs for herons and passerines, emergent vegetation for night-herons, swallows, and blackbirds, and vegetated sites like water lilies or lotus beds, flooded shrubs, or forest canopy for swimmers like Wood Ducks (Hein and Haugen 1966; Martin and Haugen 1960). Other duck species may join Wood Ducks in these roosts (Bellrose and Holm 1994). Open water is preferred by Common Goldeneyes and other rafting sea ducks (Sayler and Afton 1981). Cranes, geese, and Southern Screamers tend to use roosts as single species, but mixes of species sometimes result by virtue of the habitat features and availability. Roost sites are an essential ingredient of the home range of such species and may be related to the location of food, ideally spaced at a minimal distance. In Sandhill Cranes, such roost sites are among major habitat features considered in management recommendations (Tacha, Nesbit and Vohs 1994). Herons and egrets, blackbirds, swallows, and often gulls and dabbling ducks commonly roost as mixed species. Protection from predators and location at an energy-saving distance from foraging sites are among the obvious advantages of such roosts. There are hundreds of papers on the topic, including many studies designed to resolve the problem these flocks may create for people: aircraft strikes on landing fields, mixed flocks of blackbirds that move into urban area and "whitewash" sidewalks and streets, and concentrations of egrets and herons that kill trees through excess nutrient deposition on root systems.

One of the more interesting and controversial ideas about roosts and other flocking is the idea that members of a roost gain visual or vocal clues to the location of food resources and seek those out as they depart the roost. This theory was first applied to terrestrial granivorous finches of Africa (termed the Information-Center Hypothesis), but some workers have concluded that herons and egrets gain information and even follow successful birds to daytime feeding areas (Krebs 1978). It also has been suggested that Cattle Egrets that lag behind at roosts may be looking for signals rather than reflecting good feeding the previous day (Siegfried 1971). Other species seem to leave in flocks but disperse to different feeding sites; nevertheless, some white birds like Snowy Egrets are conspicuous and are followed by others to feeding sites (Smith 1995). In Barnacle Geese, post-roosting gatherings (i.e., pre-feeding staging) were influenced by departure time from feeding areas the previous day. Rather than being an indication of good feeding, investigators concluded that it was organization to seek out new feeding areas because of poor results earlier (Ydenberg and Prins 1984). Establishment of nesting colonies of some seabirds seems to precede the obvious need for information on feeding but does result in subsequent opportunities to do so (Buckley 1997). Clearly, these patterns are important indicators of social organization; but detecting potential causes is a complex and difficult task.

In cold climates, winter roost sites often are selected to minimize loss of body heat by use of sheltered areas and warm waters (Albright, Owen and Corr 1983, Brodsky and Weatherhead 1984). Birds also use a variety of body postures and feature postures to minimize heat loss (Wooley and Owen 1978) and, under extreme conditions, may stay in the roost rather than fly to foraging areas, presumably because heat conservation is more important than food intake for a short period (Jorde *et al.* 1984).

References

Albright, J J., Owen, R. B. Jr, and Corr, P. O. (1983). The effects of winter weather on the behavior and energy reserves of black ducks in Maine. *Transactions of the Northeast Section Wildlife Society* **40**, 118–28.

Anderson, M. G. and Titman, R, D. (1992). Spacing patterns. In *Ecology and management of breeding waterfowl*, eds. B. D. J. Batt, A. D. Afton, M. G. Anderson, C. D. Ankney, D. H. Johnson, J. A. Kadlec, and G. L. Krapu, pp. 251–89. Minneapolis, MN: University of Minnesota Press.

Anderson, W. (1965). Waterfowl production in the vicinity of gull colonies. *California Fish and Game* **51**, 5–15.

Ashmole, N. P. (1971). Seabird ecology and the marine environment. In *Avian Biology*, Vol. I, eds. D. S. Farner and J. R. King, pp. 223–86. New York: Academic Press.

Austin, O. (1961). *Birds of the world.* New York: Golden Press.

Ball, I. J., Frost, P. G. H., Siegfried, W. R., and McKinney, F. (1978). Territories and local movements of African black ducks. *Wildfowl* **29**, 61–79.

Barnard, C. J. and Thompson, D. B. A. (1985). *Gulls and plovers: the ecology and behaviour of mixed-species feeding groups.* New York: Columbia University Press.

Bellrose, F. C. (1980). *Ducks, geese and swans of North America*, 3rd edn. Washington, DC: Wildlife Management Institute.

Bellrose, F. C. and Holm, D. J. (1994). *Ecology and management of the Wood Duck.* Washington, DC: Wildlife Management Institute.

Bennetts, R. E. and Dreitz, V. J. (1997). Possible use of wading birds as beaters by Snail Kites, Boat-tailed Grackles, and Limpkins. *Wilson Bulletin* **109**, 169–73.

Bergman, R. D. and Derksen, D. V. (1977). Observations on Arctic and Red-throated Loons at Storkerson Point, Alaska. *Arctic* **30**, 41–51.

Bergman, R. D., Swain, P. W., and Weller, M. W. (1970). A comparative study of nesting Forster's and black terns. *Wilson Bulletin* **82**, 435–44.

Blackburn, T. M. and Lawton, J. H. (1994). Population abundance and body size in animal assemblages. Philosophical Transactions of the Royal Society of London, B**343**: 33–9.

Boe, J. S. (1993). Colony site selection by Eared Grebes in Minnesota. *Colonial Waterbirds* **16**, 28–38.

Breckenridge, W. J. (1956). Nesting study of wood ducks. *Journal of wildlife Management* **20**, 16–21.

Brodsky, L. M. and Weatherhead, P. J. (1984). Behavioral thermoregulation in wintering black ducks: roosting and resting. *Canadian Journal of Zoology* **62**, 1223–6.

Buckley, N. J. (1997). Spatial-concentration effects and the importance of local enhancement in the evolution of colonial breeding in seabirds. *American Naturalist* **149**, 1091–112.

Burger, J. (1974). Breeding biology and ecology of the Brown-hooded Gull in Argentina. *Auk* **91**, 601–13.

Burger, J. (1984). Grebes nesting in gull colonies: protective associations and early warning. *American Naturalist* **123**, 327–37.

Burger, J. and Gochfield. M. (1990). *The black skimmer: social dynamics of a colonial species*. New York: Columbia University Press.

Clarke, G. A. Jr (1969). Spread-wing postures in Pelecaniformes, Ciconiiformes, and Falconiformes. *Auk* 86, 136–9.

Cody, M. (1981). Habitat selection in birds: the roles of vegetation structure, competitors, and productivity. *BioScience* 31, 107–13.

Delacour, J. (1954–64). *Waterfowl of the world*, Vols. 1–4. London: Country Life.

Eldridge, J. L. (1986). Territoriality in a river specialist, the blue duck. *Wildfowl* 37, 123–35.

Elowson, A. M. (1984). Spread-wing postures and the water repellancy of feathers: a test of Rijke's hypothesis. *Auk* 101, 371–83.

Emlen, S. T., and Ambrose III, H. W. (1970). Feeding interactions of Snowy Egrets and Red-breasted Mergansers. *Auk* 87, 164–5.

Emlen, S. T., Demong, N. J., and Emlen, D. J. (1989). Experimental induction of infanticide in female Wattled Jacanas. *Auk* 106, 1–7.

Erlich, P. R., Dobkin, D. S., and Wheye, D. (1988). *The birder's handbook*. New York: Simon and Schuster.

Fjeldsa, J. (1985). Origin, evolution, and status of the avifauna of Andean wetlands. *Ornithogical Monographs* 36, 85–112.

Fredrickson, L. H. (1970). The breeding biology of American coots in Iowa. *Wilson Bulletin* 82, 445–57.

Fry, C. H., Fry, K., and Harris, A. (1996). *Kingfishers, bee-eaters and rollers*. Princeton, NJ: Princeton University Press.

Gaston, K. J. and Blackburn, T. M. (1996). Conservation implications of geographic range: body size relationships. *Conservation Biology* 10, 638–46.

Gibb, J. (1956). Food, feeding habits and territory of the Rock Pipit. *Ibis* 98, 506–30.

Gloutney, M. L. and Clark, R. G. (1997). Nest-site selection by mallards and blue-winged teal in relation to microclimate. *Auk* 114, 381–95.

Gullion, G. W. (1953). Territorial behavior in the American coot. *Condor* 55, 169–86.

Harrington, B. (1996). *The flight of the red knot*. New York: W. W. Norton.

Hein, D. and Haugen, A. O. (1966). Autumn roosting flight counts as an index to wood duck abundance. *Journal of Wildlife Management* 30, 657–68.

Hennemnann, W. W. (1982). Energetics and spread wing behavior of Anhinga in Florida. *Condor* 84, 91–6.

Hochbaum, H. A. (1944). *The Canvasback on a prairie marsh*. Washington, DC: American Wildlife Institute.

Holmes, R. T. and Pitelka, F. A. (1968). Food overlap among coexisting sandpipers on northern Alaska tundra. *Systematic Zoology* 17, 305–18.

Howard, E. (1940). *The waterhen's world*. London: Cambridge Univ. Press.

Jorde, D. G., Krapu, G. L., Crawford, R. D., and Hay, M. A. (1984). Effects of weather on habitat selection and behavior of mallards wintering in Nebraska. *Condor* 86, 258–65.

Kahl, M. P. (1964). Food ecology of the Wood Stork (*Mycteria americana*) in Florida. *Ecological Monographs* 34, 97–117.

Kiel, W. H. Jr (1970). *A release of hand-reared mallards in South Texas*. Texas A&M University Agriculture Experiment Station Publication MP968. Austin TX: University of Texas Press.

Krebs, J. R.(1978). Colonial nesting in birds, with special reference to Ciconiiformes. In Research Report No. 7: *Wading birds*, eds. A. C. Sprunt IV, J. C. Ogden, and S. Winckler, pp. 299–311. New York: National Audubon Society.

Kushlan, J. A. (1981). Resource use strategies by wading birds. *Wilson Bulletin* 93,145–63.

Kushlan, J. A., Morales, G., and Frohring, P. C. (1985). Foraging niche relations of wading birds in tropical wet savannas. *Ornithological Monographs* 36, 663–82.

Lack, D. (1968). *Ecological adaptations for breeding in birds*. London: Methuen.

LeSchack, C. R. and Hepp, G. R. (1995). Kleptoparsitism by American Coots on Gadwalls and its relationship to social dominance and food abundance. *Auk* 112, 429–35.

Mahoney, S.A. (1984). Plumage wettability of aquatic birds. *Auk* 101, l8l-95.

Martin, E. and Haugen, A. O. (1960). Seasonal changes in wood duck roosting flight habits. *Wilson Bulletin* 72, 238–43.

McFarlane, R. W. (1975). Notes on the Giant Coot (*Fulica gigantia*). *Condor* 77, 324–7.

McIntyre, J. W. (1978). Wintering behavior of common loons. *Auk* 95, 396–403.

McKinney, F. (1965). Spacing and chasing in breeding ducks. *Wildfowl* 16, 92–106.

McKinney, F. (1992). Courtship, pair formation, and signal systems. In *Avian Biology*, Vol. VIII, eds. D. S. Farner,, J. R. King, and K. C. Parkes, pp. 214–50. New York: Academic Press.

Myers, J. P., Connors, P. G., and Pitelka, F. A. (1979). Territoriality in non-breeding shorebirds. *Studies in Avian Biology* 2, 231–46.

Nickell, W. P. (1966). Common Terns nest on muskrat lodges and floating cattail mats. *Wilson Bulletin* 78, 123.

Nudds, T. D. and Bowlby, J. N. (1984). Predator–prey size relationships in North American dabbling ducks. *Canadian Journal of Zoology* 62, 2002–8.

Nudds, T. D. and Wickett, R. G. (1994). Body size and seasonal coexistence of North American dabbling ducks. *Canadian Journal of Zoology* 72, 779–82.

Nuechterlein, G. L. (1981). Information parasitism in mixed colonies of western grebes and Forster's terns. *Animal Behavior* 29, 985–9.

Orians, G. W. (1971). Ecological aspects of behavior. In *Avian Biology*, Vol. I, eds. D. S. Farner, J. R. King, and K. C. Parkes, pp. 513–46. New York: Academic Press.

Orians, G. W. (1980). *Some adaptations of marsh-nesting blackbirds*. Princeton, NJ: Princeton University Press.

Oring, L. W. (1982). Avian mating systems. In *Avian Biology*, Vol. VI, eds. D. S. Farner, J. R. King, and K. C. Parkes, pp. 1–92. New York: Academic Press.

Oring, L. W. and Knudsen, M. L. (1972). Monogamy and polyandry in the Spotted Sandpiper. *Living Bird* 11, 59–73.

Pehrsson, O. (1976). Food and feeding grounds of the Goldeneye *Bucephala clangula* (L.) on the Swedish west coast. *Ornis Scandinavica* 7, 91–112.

Pettingill, O. S. Jr (1970). *Ornithology in laboratory and field*. Minneapolis, MN: Burgess.

Picman, J. (1980). Impact of marsh wrens on reproductive strategy of Red-winged Blackbirds. *Canadian Journal of Zoology* 58, 337– 50.

Picman, J., Pribil, S. and Picman, A. K. (1996). The effect of intraspecific egg destruction on the strength of marsh wren eggs. *Auk* 113, 599–607.

Reicholf, J. (1975). Biogeographie und Okologie der wasservogel im subtropisch-
tropischen Sudamerika (with English summary). *Anzeiger der Ornithologischen
Gesellschaft in Bayern* 14, 1–69.

Ripley, S. D. (1957). Notes on the Horned Coot, *Fulica cornuta* Bonaparte. Yale
Peabody Museum of Natural History. *Postilla,* 30, 1–8.

Rohwer, F. C. (1992). Evolution of reproductive patterns. In *Ecology and management
of breeding waterfowl,* eds. B. D. J. Batt, A. D. Afton, M. G. Anderson, C. D. Ankney,
D. H. Johnson, J. A. Kadlec, and G. L. Krapu, pp. 486–539. Minneapolis, MN:
University of Minnesota Press.

Sayler, R. D. and Afton, A. D. (1981). Ecological aspects of Common Goldeneyes
Bucephala clangula wintering on the Upper Mississippi River. *Ornis Scandinavia*
12, 99–108.

Schoener, T. W. (1968). Sizes of feeding territories among birds. *Ecology* 49, 123–41.

Schreiber, R. W. and Ashmole, N. P. (1970). Sea-bird breeding seasons on Christmas
Island, Pacific Ocean. *Ibis* 112, 363–94.

Siegfried, R. (1971). Communal roosting of the Cattle Egret. *Transactions of the Royal
Society of South Africa* 39, 419–43.

Skutch, A. F. (1976). *Parent birds and their young.* Austin, TX: University of Texas
Press.

Smith, J. P. (1995). Foraging sociability of nesting wading birds (Ciconiiformes) at
Lake Okeechobee, Florida. *Wilson Bulletin* 107, 437–51.

Southwood, T. R. E. (1977). Habitat, the template for ecological strategies? *Journal
of Animal Ecology* 46, 337–65.

Stutzenbaker, C. D. (1988). *The Mottled Duck.* Austin, TX: Texas Parks and Wildlife
Department.

Svardson, G. (1949). Competition and habitat selection in birds. *Oikos* 1, 157–74.

Tacha, T. C., Nesbitt, S. A. and Vohs, P. A. Jr (1994). Sandhill crane. In *Migratory shore
and upland game bird management in North America,* eds. T. C. Tacha, & C.
Braun, pp. 77–94. Washington, DC: International Fish and Wildlife Agencies.

Turner, D. A. and Gerhart, J. (1971). "Foot-wetting" by incubating African Skimmers
Rynchops flavirostris. Ibis 113, 244.

Tyler, S. and Ormerod, S. (1994). *The dippers.* San Diego, CA: Academic Press.

Verner, J. and Engelson, G. H. (1970). Territories, multiple nest building, and poly-
gyny in the Long-billed Marsh Wren. *Auk* 87, 557–67.

Verner, J. and Willson, M. F. (1966). The influence of habitat on mating systems of
North American passerine birds. *Ecology* 47, 143–7.

Vuilleumier, F. (1975). Zoogeography. In *Avian Biology,* Vol. V, eds. D. S. Farner,, J. R.
King, and K. C. Parkes, pp. 421–96. New York: Academic Press.

Weller, M. W. (1964). Distribution and species relationships. In *The waterfowl of the
world,* Vol. IV., ed. J. Delacour, pp. 108–20. London: Country Life.

Weller, M. W. (1968). The breeding biology of the parasitic black-headed duck. *Living
Bird* 7, 169–207.

Weller, M. W. (1975). Migratory waterfowl: a hemispheric perspective. In *Symposium
on wildlife and its environment in the America's. Publiciones Biologicas Instituto de
Investigacions Cienificas, Universitad Autonoma de Nuevo Leon* 1, 89–130.

Weller, M. W. (1979). Density and habitat relationships of blue-winged teal nesting
in Northwestern Iowa. *Journal of Wildlife Management* 43, 367–74.

Weller, M. W. (1980). *The island waterfowl.* Ames, IA: Iowa State University Press.

Weller, M. W. (1995). Use of two waterbird guilds as evaluation tools for the Kissimmee River restoration. *Restoration Ecology* 3, 211–24.

Weller, M. W. and Spatcher, C. E. (1965). *Role of habitat in the distribution and abundance of marsh birds.* Iowa State University Agriculture & Home Economics Experiment Station Special Report, 43. Ames IA: Iowa State University Press.

Wittenberger, J. F. and Hunt, G. L. Jr (1985). The adaptive significance of coloniality in birds. In *Avian Biology*, Vol. VIII, eds. D. S. Farner,, J. R. King, & K. C. Parkes, pp. 1–78. New York: Academic Press.

Wooley, J. B. Jr and Owen, R. B. (1978). Energy costs of activity and daily energy expenditures in the black duck. *Journal of Wildlife Management* 42, 739–45.

Ydenberg, R. C. and Prins, H. H. Th. (1984). Why do birds roost communally. In *Coastal waders and wildfowl in winter.* eds. P. R. Evans, J. D. Goss-Custard, & W. G. Hale, pp. 123–39. Cambridge: Cambridge University Press.

Zimmerman, D. A., Turner, D. A., and Pearson, D. J. (1996). *Birds of Kenya and northern Tanzania.* Princeton, NJ: Princeton University Press.

8

Spatial and structural patterns

Wetlands are especially known for their abundance of some species or groups of birds as well as for their diversity of species not found elsewhere. The major reason for this abundance presumably is the availability of resources produced by the great primary productivity of wetlands. The diversity of species undoubtedly is linked to diversity of resources produced in the many micro-habitats resulting from water-depth gradients and fluctuations. The range of spatial scales we can examine was outlined above, but I would like here to focus on factors that influence bird use within and among wetlands. Although the emphasis is on geographical, and physical or structural patterns in wetlands, this includes structural aspects of live components such as plants (Table 8.1). Plants vary in **life form** or physical stature, which, with water distribution, creates visual patterns useful to us as indicators of wetland types and conspicuous to birds as indicators of various essential resources. We will first consider the range from **homogeneity** to **heterogeneity** that can exist in certain wetland types, forming within-wetland variation of habitats (microhabitats), and then consider larger-scale relationships among wetlands or wetland complexes or regimes, which can best be appreciated by first understanding the patterns found in individual wetlands.

8.1 Vegetation life form

Wetland types differ in the height, diameter, and robustness of dominant plants as well as in the range of plant species diversity, which with variation in water regimes can result in highly variable and sometime quite complex patterns of habitat diversity. Birds seem to recognize structural features of wetland plants as well as their values for food, making vegetation structure convenient for making generalizations about bird/vegetation relationships, and thereby assessing the potential value to birds.

From the completely aquatic to the more terrestrial, these plant life forms result from adaptation to the presence, depth, duration, seasonality, temperature, and chemistry of water. These form different habitats from which a species can chose its microhabitat for particular activities. Various bird species

Table 8.1 *Wetland structural and physical features that influence habitat classification and segregation of bird species and groups*

Basin size

Basin shoreline configuration (development)

Basin substrate materials (clay, silt, sand, etc.)

Basin shoreline slope (rate of change)

Water depth and variability

Water clarity, salinity, icing, etc.

Water column structure (plant layering and distribution underwater)

Hydrologic influences: current, waves, and tides

River gradient and volume

Pools or openings in vegetation (size, dispersion)

Vegetation zonation and other distribution patterns

Cover–water ratios

Plant density

Plant height

Plant sturdiness or robustness

Plant aerial coverage based on life form: graminoid, herbaceous, woody, bush, tree, etc.

Vertical layering or strata

have evolved different preferences, which suggests how habitat diversity can result in habitat sharing of different species in a small area (Weller and Spatcher 1965). Such zones can change rapidly over time through catastrophic events such as flooding or fire, which cause plant mortality and also create new plant growth sites through soil exposure or sedimentation. Subsequent invasion occurs via underground rhizomes, mass seedings, or major germination events from the seed bank. Even without catastrophic events, more gradual biotic change is common. These dynamic patterns result in changes in availability of resources as well as in structure, potentially attracting different bird species at different times.

The most aquatic plants of shallow wetlands are those that are **submersed** but often rooted at some stage and are, therefore, associated either with shallow waters or with the littoral zone of lakes or ponds where light penetration is greatest. These often have filiform leaves (e.g., milfoils and pondweeds) that are able to capture light underwater more effectively. Some submersed species may survive floating in the water column when they are torn free from their roots; still other submergents are unrooted. Different species of submergent also are specialized for different depths and soil substrate and form zonal patterns that vary in value to different bird species. Dense mats can inhibit use by diving birds; others can walk on and hunt from such dense mats of surfacing submergents.

Plants that float on the water's surface (**floating plants,** e.g., Lesser Duckweed and Water-hyacinth) or occasionally in the water column (Star Duckweed) and take their nutrients directly from the water via suspended roots or the osmotic processes of individual cells are less restricted in distribution in a wetland – when water is present. Larger species of floating vegetation will support the weight of small walking birds like jacanas, Purple Gallinules, or small egrets and bitterns, but such mats often inhibit swimming by larger birds.

Often found in water of variable depths such as riverine backwater oxbows and swamps, **floating-leaf** or **pad plants** (e.g., water-lilies) can flourish in fluctuating or turbid water because they send up long stalks from often huge and buried tubers (which are in themselves a valuable food for vertebrates). There are intermediate plant growth strategies of course, and the Yellow Lotus is one that may start its growth as a pad plant and mature as a robust emergent as water support declines and the stalk strengthens.

Much if not most of the great productivity of wetlands comes in the form of rooted herbaceous **emergent plants,** which live in the air/sunlight medium that is ideal for photosynthesis but also can tap the nutrients of the basin soil. Examples are cattail, sedges, bulrushes, rushes, spike-rushes, and the more water-tolerant grasses such as cordgrasses in shallow water or damp soil.

Although they often grow by emerging from the water, **woody plants** like scrub, shrubs, and trees are separated from herbaceous emergent plants in classifications because of the dramatic difference in physical structure and mass. In North America, they include water-tolerant shrubs like Buttonbush, which germinate in shallow water or moist soil but survive as flooded plants for many years. Other woody shrubs include alder and shrub willow which occur along the shoreline slopes of basin wetlands and in stream riparian zone sites where periodic water enables germination and survival. Similar scrub or shrub species occur in slope wetlands such as the seeps from snow fields at high altitude, even though there is rarely standing water. Several species of small tree like willows are common in wet sites around the world, being very successful mudflat pioneers. Many species of larger trees establish in wet sites of stream riparian zones or floodplains that are seasonally flooded (cottonwood), nearly continuously flooded (some oaks and gums), or continuously flooded (Bald-cypress). Mangroves of many species dominate warm-climate tidal wetlands around the world and vary from single shrubs to massive forests.

8.2 Horizontal patterns

Wetland plant variety produces variation in habitat structure through species' differences in heights and density, and through patterns of plant establishment, competition, and interaction. In basin and fringe wetlands, such patterns reflect plant adaptation to different and often varying water depths in the

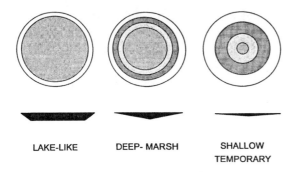

LAKE-LIKE   DEEP- MARSH   SHALLOW
                           TEMPORARY

Figure 8.1. Typical vegetation zonation in relation to water depth in three major basin wetland types or stages. These can be very dynamic over periods of several years, with variable water depth and hydroperiod.

wetland or at the wetland edge, producing concentric patterns or **zonation** (Fig. 8.1)(Spence 1982). Additionally, basin configuration, soil texture, prior vegetation history, water quality, herbivores, and many other factors influence such zones. Bird distribution patterns reflect these microhabitat features and range from single to multiple species distributed in varying patterns from homogeneous to heterogeneous. Often, patterns change in the same wetland over time and water regime. Such distribution patterns have been mapped, but less commonly have been related specifically to the plant species or to the attributes of the vegetation and water regimes that produce the pattern. Some birds favor the interface or edges between conspicuous life forms (Beecher 1942, Weller and Spatcher 1965); others use tall and fairly uniform vegetation (Marsh Wrens, Nearctic blackbirds, herons); some use short vegetation (Sedge Wrens), and some walkers use dense (bitterns) or mostly sparse vegetation (coots and grebes). Some use dead vegetation from prior growing seasons, and others use mainly new growth. In all cases, we assume that water depths and food resources also are key issues in microhabitat selection. Foresters and urban landscape workers have perhaps done a better job of identifying the association between birds and vegetation structure (DeGraaf 1986) than have wetland scientists.

8.3 Vertical structure

Perhaps the most obvious structural habitat component that influences bird use is vertical structure, first recognized by forest ecologists. Descriptive measurements include not only height and number of vegetation layers, but foliage volume and leaf form (MacArthur and Wilson 1967). Such factors are equally important for waterbirds, regardless of what may seem like minor height differences in emergent marshes; in addition, waterbirds confront another layer, the water column, that terrestrial birds do not use. Moreover, this is complicated by the fact that food resources may be at any height above the water, float on the water's surface, be distributed at any level within the water column, be mainly on the surface of the wetland bottom (e.g., epibenthic

Spatial and structural patterns

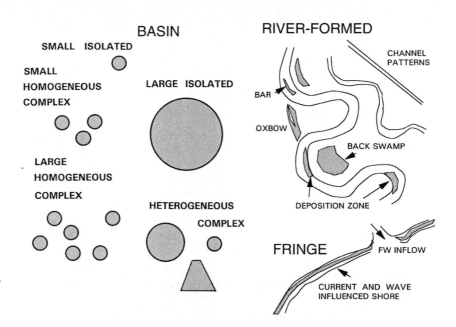

Figure 8.2. Patterns of wetland diversity and potential importance in bird use and distribution related to geomorphology and formative processes. FW, freshwater.

organisms), or may be buried in the substrate (e.g., endobenthic). Hence, there are various rich resources that induce not only morphological and anatomical adaptations but also behavioral specialization. Waterbirds that nest in burrows (kingfishers, Crab-Plovers), crevices (dipper and Torrent Duck), tree holes (Tree Swallow, Wood Duck, Bufflehead, and several mergansers), tree branches (Marbled Murrelet), or cliff surface (several geese and raptors) deal directly with the vertical layer above the water's surface.

8.4 Wetland size

Some wetland diversity often is a product of size. It has been known for many years that increasing the size of sample plots in plant communities results in increased number of species (i.e., the species–area curve). Increased island size results in increased number of vertebrates, including birds (Duebbert 1982, He and Legendre 1996, MacArthur and Wilson 1967), and wetlands seem to function in the same way (Fig. 8.2). Increased species richness with wetland size has been reported for ducks in small prairie and forested wetlands (Nudds 1992), forested bog wetlands in Maine (Gibbs *et al.* 1991), (Figs. 8.3 and 8.4), diverse waterbirds in lakes and smaller wetlands in Italy (Celada and Boglianai 1993), and Finland (Lampolahti and Nuotio 1993), and estuarine tidal wetlands (Craig and Beal 1992). Hirano and Higuchi (1988) reported that broader riparian areas along large streams also demonstrated increased bird richness over smaller streams in winter, mainly because of the extensive habitats for waders produced by expanded bars and mudflats (Fig. 8.5). The influence of river

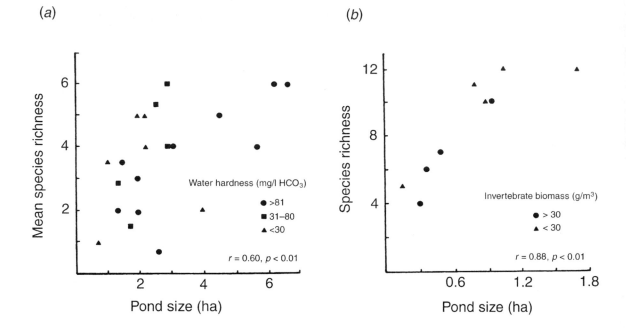

(a)

(b)

Figure 8.3. Relationship of species richness to pond size for ducks in small forested beaver ponds of Eastern Canada compared with the more productive prairie basin wetlands. Data recorded by water hardness (*a*) and invertebrate biomass (*b*). (From Nudds 1992, with permission.)

width on dipper populations is less clear, but lower stream gradients seem to be favored in some areas, inferring use of wider streams with more diverse habitats (Tyler and Ormerod 1994). Studies of streamside or riparian habitats suggests that broad areas hold more species of neotropical migrants than more narrow ones (Hodges and Krementz 1996, Keller, Robbins and Hatfield 1993), resulting in the recommendation for a 100 m minimal buffer zone, which is especially important for these neotropical migrants.

Within-wetland vegetation heterogeneity clearly influences bird species richness and the abundance of breeding birds in the vegetation of freshwater marshes (Beecher 1942, Kaminski and Prince 1984, Weller and Spatcher 1965). In one study of small tidal and nontidal wetlands (Craig and Beal 1992), the influence of size exceeded that of heterogeneity for "breeders," but other bird "users" showed a greater but not major relationship. An important observation for conservation of species' diversity was that larger wetland units were most important in providing habitat for those species that occurred in low density, a conclusion in agreement with observations by Burger *et al.* (1985) for tidal salt marsh, where diversity was less than in impoundments.

8.5 Wetland diversity

Because many birds use several wetlands to accomplish goals like pairing or rearing young, and because diversity of wetlands creates habitat diversity

Spatial and structural patterns

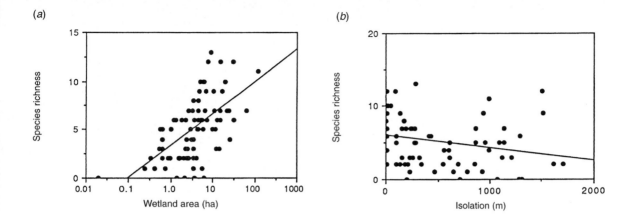

(a)

Species richness — Wetland area (ha)

(b)

Species richness — Isolation (m)

Figure 8.4. The relationship of species richness to area (*a*) and isolation pattern (*b*) of diverse waterbirds in 87 Maine wetlands (Gibbs *et al.* 1991, with permission).

attractive to different species of birds, **wetland complexes** also show species–area patterns (Fig. 8.6). The relationship and evaluation of wetlands within such a complex is difficult to study, but several workers have demonstrated the correlation of waterfowl abundance and species richness with wetland heterogeneity (Brown and Dinsmore 1986, 1991, Flake 1979, Patterson 1976). Additionally, there is some evidence to indicate that more isolated wetlands have fewer species (Fig. 8.4*b*). These concepts are based on studies of oceanic islands and often consider nonmigratory species or populations. Wetlands may be more dynamic habitats owing to water regimes, and often they are used only seasonally or periodically by strongly migratory or nomadic birds; as a result, they demonstrate a more dynamic pattern of bird use over time. Therefore, wetland bird diversity and community composition of a landscape region may be strongly influenced by the diversity of wetland sizes present and by the pattern of wetness and dryness at a given time. If wetlands of all sizes are at their extremes of wetness or dryness, composition would be drastically changed over intermediate stages – and wetter is not usually better!

From a conservation perspective, the importance of wetland complexes is that they not only provide a variety of habitats for a variety of species but that those habitats influenced by water and vegetation regimes change in pattern, which will make some units within the complex useful to migrant birds like shorebirds when others may be unavailable or unattractive (Skagen and Knopf 1994). Moreover, the loss of smaller wetlands impinges on different species than will the loss of a larger wetland because of the more aquatic bird fauna of the latter and the more terrestrial species of the former (Weller 1979). The temporal aspects of these periodic pulses will be considered below.

(a)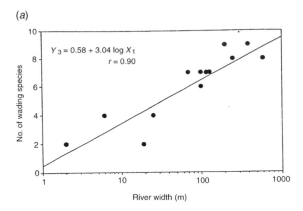

(b)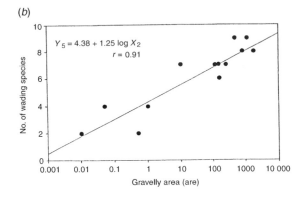

8.6 Wetland configuration

Shoreline configuration is one of the more conspicuous features that would seem to influence bird use, especially during the breeding season when aggressive behavior is prominent. Meaningful parameters of such edges are difficult to relate to birds and other mobile vertebrates, but most workers use correlations with the **shoreline index,** a ratio of shoreline length to water area. Based on studies such as those in Sweden, dabbling and bay ducks and coots show a significant relationship to shoreline index (Fig. 8.7), whereas Great-crested Grebes do not (Nilsson 1978). Similar relationships have been observed by other workers (see Kaminski and Weller 1992) and have been demonstrated experimentally (Kaminski and Prince 1984).

Figure 8.5. Relationship between species richness of wading birds and river width (*a*) and extent of gravel areas (*b*) (1 are = 100 m^2) (Hirano and Higuchi 1988, with permission).

8.7 General habitat patterns and descriptors

By using some of the above structural habitat patterns as descriptors of bird habitats (or microhabitats), it should be more possible to categorize, describe, and perhaps quantify species, distribution and segregation within a bird community that is using a heterogeneous wetland. As used here, the term habitat pattern infers those features of the bird's immediate habitat at a scale of the bird's visual image needed to select its living place for a specific time and function. Vegetation structure and the water features are the only practical parameters because these also are the images that humans recognize. What is uncertain is the scale of vegetation structure and, as mentioned above, that probably is a hierarchical assessment from large-scale plant communities like biomes or biogeographic regions, through wetland regions and wetland complexes, to individual wetlands, and, finally, to plant communities and water depth within the wetland. It would seem logical that wetland birds would home to a single wetland within a biogeographical region, and this is a more practical approach to habitat description.

Spatial and structural patterns

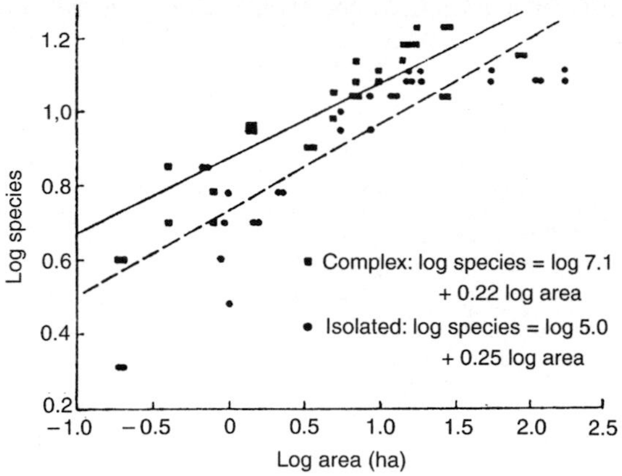

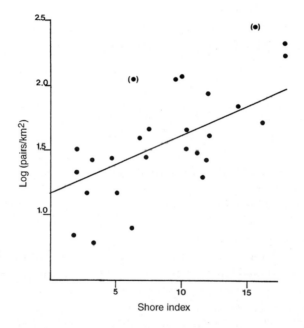

Figure 8.6. Relationship between species richness and wetland size for 15 individual wetlands and 15 wetland complexes. (From Brown and Dinsmore (1986) with permission © The Wildlife Society.)

Complex: log species = log 7.1 + 0.22 log area

Isolated: log species = log 5.0 + 0.25 log area

Figure 8.7. Relationship between the density of breeding ducks and shoreline development in some eutrophic Swedish lakes (Nilsson 1978, with permission).

Tree, shrub, bush, forb, and graminoid forms all influence whether the foliage layer is high above the water or low, the robustness and obscurity produced by the size of leaves, stems, stalks, and trunks, and the dispersion and density of vegetation in relation to water (semiopen or densely vegetated). These factors influence the birds' ability to walk or swim through the vegetation for food or cover, to build nests in it, to sun, or to see its neighbors. These are difficult parameters to identify and measure, and subsequently a statistical correlation of bird use with specific vegetation characters is necessary. Moreover, these parameters are often highly variable by time and place. This type of research is becoming more common, and each attempt identifies some

Table 8.2 *Wetland habitat patterns that serve as descriptors of wetland types and habitats for bird assemblages*

Deep, open-water

Submerged aquatic beds

Floating vegetation and pad plants

Robust and persistent herbaceous emergents (graminoids)

Nonpersistent, meadow-like emergents

Bog mats and organic islands

Shrubs

Forested wetlands

Unvegetated shallow water and mudflats

Tidal flats and salt marsh shores

Riverine sandbars, islands, and depositional areas

Boulders, rock substrate and other hard-shores

Constructed pond, lake, and reservoir

Riverine shore, substrate, and pools

associations that allow better understanding, potential modeling, and possible prediction of habitat quality and use (e.g., Colwell and Dodd 1997).

Water openness seems very important, but its relative impact will vary with the season and the stage of the bird's life-cycle. Small walking birds that also swim (e.g., gallinules) often swim to plant-covered water and then walk. Dedicated swimmers tend to avoid plant-covered waters unless seeking out remote nesting sites or protection from a predator. Shorebirds like the peeps and "clam diggers" prefer totally bare but suitable wet mudflats. Others like yellowlegs or Greenshanks with long legs prefer sheetwater with sparse vegetation. They are replaced by plovers, or terrestrial birds like pipits and doves as the waters dry; sandbars and sandflats attract plovers, terns, and skimmers. However, only the more obvious patterns have been documented, and the factors that influence behavioral variation are poorly known.

Thus, we can classify bird habitats using a combination of water presence and depth, wetland shoreline configuration, water hydraulics, and vegetation, which all create conditions suitable for common bird foods (Chapter 5). How useful these descriptors are depends upon our goals and precision.

To demonstrate how birds typically are associated with these wetland habitats (Table 8.2), I list below some common North American birds typically found in these habitats (modified from Weller 1996). The danger of descriptions like these is that they may vary in character by region and may change seasonally. This list is neither complete nor final but rather a structure or outline useful for regional summaries and comparisons. Workers in various bird groups have refined these categories based both on their knowledge of how species segregate to feed or nest and on the practicality of field

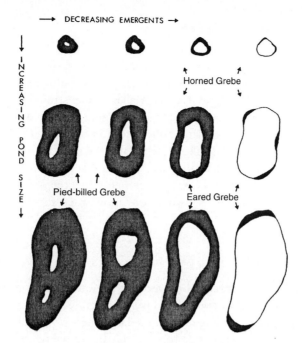

DECREASING EMERGENTS →

INCREASING POND SIZE

Horned Grebe

Pied-billed Grebe

Eared Grebe

Figure 8.8. Selection of wetland vegetation patterns by three species of North American grebes, as related to pond size and vegetation cover (Faaborg 1976, with permission).

identification and description. For example, Withers and Chapman (1993) identified at least three microhabitats to segregate shorebird survey data on a saline mudflat. Harrington (1996) used seven categories to describe habitat use by shorebirds in the Cheyenne Bottoms of Kansas, grading from open water to dry and vegetated upland. But in addition to making the description fit the study, we do need to enhance habitat descriptors and incorporate them in all bird guides and regional works.

8.7.1 BIRDS OF DEEP OPEN WATER

Most birds of deeper, open water with lake-line shorelines are divers that feed on mobile fish, large invertebrates, and occasionally amphibians such as salamanders. These include loons, pelicans, cormorants, certain grebes, and fish-eating raptors like Ospreys and Bald Eagles (Esler 1992). Some are benthic divers that feed on clams, worms, or insect larvae, such as Lesser Scaup, Canvasback or Redhead ducks (Bartonek and Hickey 1969, Bergman 1973, Woodin and Swanson 1989). A few species, such as Northern Shoveler, are planktonic feeders and take tiny crustaceans by straining though special bill morphology. They tend to feed in shallow, rich waters in groups or with other birds that stir-up waters and make prey vulnerable. During nesting periods, these species are more closely linked to nesting cover or islands, as reflected in wetland choice by several species of small grebes (Faaborg 1976,) (Fig. 8.8).

8.7.2 BIRDS OF SUBMERGED AQUATIC BEDS

Birds that use submergent plant beds during the nonbreeding period are those like American Coots, swans, Gadwall, American Wigeon, or Canvasbacks and Redheads, which feed on the foliage or tubers of plants such as pondweeds, wigeongrass, and seagrasses. Blue-winged Teal and Mallards may strain out floating seeds or surface invertebrates, strip seed from emerging heads, pluck off associated invertebrates such as snails, and bottom feed in more shallow water (Swanson and Meyer 1977). Pied-billed Grebes also feed among the vegetation, seemingly able to move easily through the dense vegetation to capture small fish and large invertebrates that take cover there (Esler 1992). The little Snowy Egret may walk on dense mats of such vegetation or use them as stationary feeding perches.

8.7.3 BIRDS OF FLOATING VEGETATION AND PAD PLANTS

Small floating vegetation like duckweed, wolfweed, and Water-fern may attract or inhibit use by waterbirds. Most birds find invertebrates are abundant there and it is, therefore, a prime feeding area. Some larger herbivores like Southern Screamers may eat duckweed and probably take in animal matter incidentally. When combined with submergent vegetation, which may trap windblown floating plants, floating vegetation can be a serious problem for small swimming birds and especially for young birds because it inhibits rapid escape by swimming or diving. Small floating-leaf plants like Watershield and some pondweeds typically are in low density and cover because of underwater foliage, and they are often rich in near-surface seed heads attractive to various swimming seed-eaters. Water lilies and other large-leaved pad plants may support the weight of jacanas, downy ducklings, and blackbirds or other passerines pursuing aquatic invertebrates. These and taller plants like lotus form excellent escape and roosting cover for many swimming birds.

8.7.4 BIRDS OF ROBUST OR PERSISTENT HERBACEOUS EMERGENTS

Emergent vegetation has the greatest concentration of species, presumably because of its diverse food resources and structural opportunities of cover for different species and different functions during individual life cycles. Use of emergents is especially prominent during the nesting season, when nests of different species may be at various sites. Cover–water patterns are important because species often favor edges, and swimming birds move between patches of vegetation via water channels. Moreover, some birds require sizable areas for taking flight. Therefore, dense stands of single species of plant are less attractive to most species of birds (Kaminski and Prince 1984, Murkin, Kaninski and Titman 1982, Weller and Fredrickson 1974, Weller and Spatcher 1965) and maintenance of balanced cover–water interspersion (i.e., "hemi-marsh") is a management target used by wildlife management agencies. Invertebrates also may be more abundant among these emergents in some seasons (Murkin, Murkin and Titman 1992) and attract birds that use such foods in preparation for egg laying.

Persistent emergents are favored by conspicuous blackbirds in North America, such as Yellow-headed Blackbirds and several other blackbirds in western marshes and Red-winged Blackbirds and Boat-tailed Grackles in the midwest and eastern USA (Burger 1985, Orians 1980). All build cup-shaped nests of local plant materials and position the nest below the tips of the vegetation for cover and wind protection. Sedge Wrens nest and feed in dense grasses and sedges of shallow waters, whereas Marsh Wrens use taller vegetation in more permanent water. Both build ball-shaped nests above the flooding zone and find suitable invertebrate food at water level as well as in the upright vegetation. Walking rails like the Purple Gallinule, Virginia Rail, Sora and King Rail grade in preference from shallow water to almost dry ground (Weller and Spatcher 1965). However, all are adaptable, and their precise nest site and feeding location in relation to vegetation structure seems subject to variation in water regime, vegetation density, food availability, season, and perhaps local population traditions. Herons, egrets, bitterns and ibises may nest among or on top of robust emergents like cattail or California Bullrush, and some, like bitterns, also feed there during nesting. Such stands of vegetation are excellent waiting sites for stationary feeders like Least Bitterns or Tricolor Herons. The Canada Goose and several duck species may nest on lodges built of emergent vegetation by muskrats or beaver.

Some birds favor the protection of emergents compared with extensive open-water areas that may be influenced by winds or waves but still use the resources of the shallow emergents. Among these are diving ducks like the Ring-necked Duck and Ruddy Duck, which may nest in emergents but feed in the dense submergents often found in these pools. Because they are excellent swimmers and divers, they escape predators by diving and move to adjacent open water to take flight. The swimmers among the rail family, including coots, using deeper water and Moorhens of more shallow water are common in these pools. Black Terns feed on invertebrates at the surface and nest on small mats in such openings. Ibises and large rails use such pools for feeding, and many ducks rear their broods in such sites where cover is adjacent to food.

Salt marshes may differ in species and in response of birds to tidal water regimes, above, as indicated but their patterns of microhabitat selection and foraging strategies are similar.

8.7.5 BIRDS OF NONPERSISTENT, MEADOW-LIKE EMERGENTS

Seasonally flooded meadows of shorter grasses, spikerush, or annual forbs and shoreline emergent zones of similar vegetation structure are used by diverse ground-feeding and walking birds – depending on water depth. During wetter

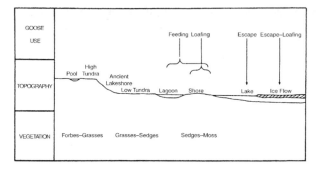

Figure 8.9. Habitat use of various geomorphic zones and ice floes by Black Brant and other nonbreeding geese on the Alaskan Coastal Plain (after Derksen, Eldridge and Weller (1982) with permission).

periods, long-legged waders feed there, such as Black-necked Stilts and avocets in saline areas, and dowitchers, yellowlegs, and sometimes ibises and egrets if large invertebrate foods are present in temperate areas. Grazing waterfowl such as White-fronted Geese, Brant, or American Wigeon forage on new shoots resulting from spring flooding of meadow or tundra vegetation (Fig. 8.9). Seed-eating finches, plovers, and shorebirds also feed there and may nest in the developing cover as the water recedes.

Birds that prefer persistent emergent plants and shorter and less robust nonpersistent vegetation also are conspicuous at the shallow edges of wetlands or the perimeter of large islands within a wetland. These areas tend to be dominated by walkers rather than swimmers, including King and Clapper Rails, American Bitterns, Snowy Egrets, and Little Blue Herons. Red-winged Blackbirds prefer the edges and are less abundant in deeper water. Some species spend as much time in the adjacent uplands as at the wetland edge, such as aerial foragers like Willow Flycatchers and water or ground feeders like Swamp Sparrows and Northern Harrier (McCabe 1991, Weller and Spatcher 1965). Small trees and shrubs on unflooded upland sites are used for song perches or nest sites. Tree Swallows and Eastern Kingbirds may nest in trees over water near the shore. Other species find concentrations of important foods like insects, crayfish, and amphibians in this moist interface.

8.7.6 BIRDS OF BOG MATS AND ORGANIC ISLANDS

Buoyant organic and living vegetation forms mats where other vegetation becomes established. Moss, sedge and grassy tussock mats create areas of low life form in forests; these tend to become higher and drier over time and form sites suitable for shrubs and even trees. But these areas change slowly and provide different habitats for birds that are characteristic of boreal areas and often have greater species richness than surrounding forested habitats (Damman and French 1987). Vegetation structure and diversity most affect bird species richness, and species' use is influenced by latitude and tree composition. Taxa represented include several ducks common mainly to those areas and a variety of songbirds (Fig. 8.10), as well as one unique shorebird, the Wilson's Snipe (Tuck 1972). Bogs and peat mats also are common in south-

Spatial and structural patterns

(a)

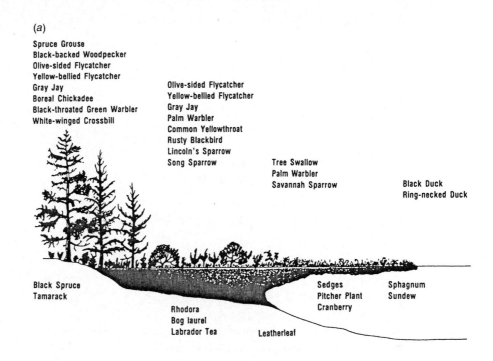

Spruce Grouse
Black-backed Woodpecker
Olive-sided Flycatcher
Yellow-bellied Flycatcher
Gray Jay
Boreal Chickadee
Black-throated Green Warbler
White-winged Crossbill

Olive-sided Flycatcher
Yellow-bellied Flycatcher
Gray Jay
Palm Warbler
Common Yellowthroat
Rusty Blackbird
Lincoln's Sparrow
Song Sparrow

Tree Swallow
Palm Warbler
Savannah Sparrow

Black Duck
Ring-necked Duck

Black Spruce
Tamarack

Rhodora
Bog laurel
Labrador Tea

Leatherleaf

Sedges
Pitcher Plant
Cranberry

Sphagnum
Sundew

(b)

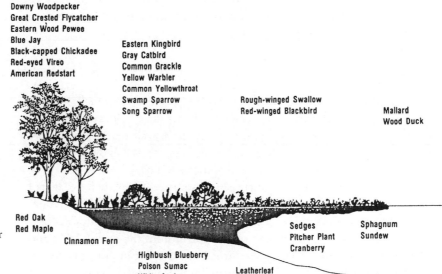

Ruffed Grouse
Downy Woodpecker
Great Crested Flycatcher
Eastern Wood Pewee
Blue Jay
Black-capped Chickadee
Red-eyed Vireo
American Redstart

Eastern Kingbird
Gray Catbird
Common Grackle
Yellow Warbler
Common Yellowthroat
Swamp Sparrow
Song Sparrow

Rough-winged Swallow
Red-winged Blackbird

Mallard
Wood Duck

Red Oak
Red Maple

Cinnamon Fern

Highbush Blueberry
Poison Sumac
White Azalea

Leatherleaf

Sedges
Pitcher Plant
Cranberry

Sphagnum
Sundew

Figure 8.10. Distribution patterns of lakeshore bog-mat wetlands and adjacent shrub and forest birds in two northeastern North American forest zones: (a) boreal and coastal spruce–fir zone; (b) Appalachian oak zone (Damman and French 1987).

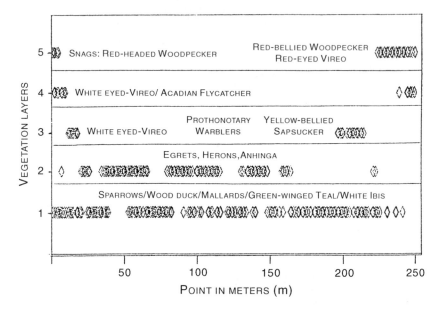

Figure 8.11. Bisect of an East Texas shrub swamp and adjacent bottomland hardwoods showing the distribution of bird species based on direct counts along transects according to five vegetation strata (Weller 1988, with permission).

eastern slope and basin wetlands where plant production exceeds decomposition partly because of saturation and reduced oxygen; these areas are best known for their pitcher plants. Glade areas in Georgia and Florida have even more extensive mats of grasses and other herbaceous plants that are used for nesting, feeding, and resting by rails, coots, and other walkers and by some songbirds, which feed on the mat or from aerial perches. Forest communities in the tropics also have bog areas, but bird species seem to be less specific to those sites, perhaps because of small size.

8.7.7 BIRDS OF SHRUB AND FORESTED WETLANDS

Wetland shrubs occur around a variety of wetland types in the seasonally flooded zone or in saturated soils in riparian or alpine habitats. Streamsides or drainages, where wetland vegetation is apparent or where terrestrial vegetation is more abundant and diverse owing to the presence of water, are well known as important riparian areas for birds (Hehnke and Stone 1978, Taylor 1986, Wauer 1977). Tree- or shrub-nesting passerines and flycatchers are prominent there because of the availability of foods from the stream and from the several layers of vegetation.

Water-induced shrubs also occur as extensive stands or patches in cold-temperate climates at high altitude or latitude around the world. Wild and domesticated herbivores browse these shrubs, and they are important habitats for Willow Flycatchers, Yellow Warblers, and numerous other passerines that exploit insect life of wetland edges or patches. More water-tolerant shrubs like Buttonbush and certain species of willow and alder can survive flooded conditions even during the growing season and are among the plant dominants in areas termed shrub–scrub swamps (Fig. 8.11). Such habitats often are used by

Spatial and structural patterns

herons and egrets as well as by passerines for nest sites. These plant zones commonly are distributed along stream banks, especially in western areas where extensive vegetation is present only because of the increased water table associated with streams.

Forested wetlands include bottomland hardwoods that are seasonally flooded during the winter dormant period. These are characteristic of backwaters of floodplains along major southeastern river systems, or of naturally impounded and trapped waters that are semipermanent and form true swamps. Birds of larger wetland trees commonly include not only the herons, which may nest at the tops of dead trees in or near water, but also woodpeckers, which make and use cavities. Large cavities in large trees may later be used by Wood Ducks and Hooded Mergansers, whereas Prothonotary Warblers and Tree Swallows use smaller holes in old snags. Such dead trees and snags must be preserved if the species are to be successful in the area, but all these species will respond to artificial nest boxes, where the expense and effort is justified. Understory plants like plume grasses and cane or bamboo form dense habitats used by Swainson's Warbler and Bachman's Warbler (which is probably now extinct); this is now a rare microhabitat because of the intense use of forested areas. Open areas are used by sparrows and wrens favoring moist and seed-rich habitats with lower herbaceous cover. As can be seen, diversity of birds here is strongly influenced by forest diversity and successional stage (Buffington *et al.* 1997, Swift, Larson and DeGraaf 1984.)

8.7.8 BIRDS OF UNVEGETATED SHALLOW WATER, MUDFLATS, AND SANDBARS

Shallow sheetwater over soil, sand, or other nonvegetated or sparsely vegetated substrate may result from rainfall, stream, or tidal flooding of mudflats and shorelines. Plowed fields and other agricultural areas also may be flooded under such circumstances, which often makes available a rich supply of foods. Open-water areas within wetlands form similar new habitats for edge birds when water levels decline below the emergent and submergent zones and expose bare basin bottoms. These offer special habitats for American Avocets, ibis and teal, which specialize in feeding in shallow water, usually by feel or straining rather than sight. Waders often walk such flats in search of fish, amphibians, reptiles, or large invertebrates in the water. Such sheetwater areas may be used also during migration or in wintering areas as overnight roosts by cranes and geese; protection of these habitats is a vital management approach to capitalizing on waste grain and other adjacent food resources attractive to such granivores (Hobaugh 1984).

Mudflats are moist and usually unvegetated areas generally not appreciated by those who do not recognize the abundance and distribution of invertebrates that occur in these habitats and the role the invertebrates play as detritivores in the cycling of nutrients within wetland systems. Thus, the mudflats offer a unique habitat for carnivorous foragers like Killdeers and numerous other plovers and sandpipers (Capen and Low 1980). More fluid mud is the favorite habitat for a number of groups of tiny crustaceans, nematodes and annelids, which are eaten especially by Green-winged Teal, a species often termed "mudder" for its walking and sifting behavior in soupy mud, and by shorebirds.

8.7.9 TIDAL FLATS

Tidal flats and shores differ from mudflats in basin wetlands because of water hydraulics, forming areas that are flooded periodically by celestial or wind tides. Therefore, some areas are available for bird use for some types of feeding only part of the time because of fluctuations in water level. Moreover, when nesting occurs in these areas, invasion by ground predators may be possible. These sites also differ in soil texture, varying from clay and silt to sand and shell deposits, which influences water drainage rates and benthic organisms and therefore, the bird species that use the habitat. Sandy or gravel (shingle) shores are common in specific areas because particle separation is a product of wave and current action on suspended or substrate materials. Therefore, mudflats commonly occur near river mouths in estuaries, with larger sand particles being deposited and regularly moved by active surf or tidal action (Britton and Morton 1989). Wind also plays a role in distribution of sand and influences resting sites, drinking areas, and even endangers bird nests. Birds that use sandy shores regularly include the well-named Sanderlings, several peeps, Red Knots, Ruddy Turnstones, and oystercatchers for feeding on clams and other burrowing invertebrates, and various gulls, terns, and skimmers for resting and nesting (Burger *et al.* 1977, Whitlack 1982). Zonation of bird use is a common product of water depth, wave forces, and food distribution (Whitlack 1982), and similar diversity occurs in salt marshes and other tidal plant communities (Daiber 1982) that develop in more stablized depositional areas. Many species forage microsites in tidal areas based on consistency of the substrate and move seaward with the ebbing and shoreward with the rising tide.

8.7.10 EMERGENT SALT MARSH

Herbaceous saltmarsh vegetation is common on stabilized sediments where seeds or rhizomes persist, forming parallel zonation similar to basin wetlands. This grades into shrubs in some areas, but this zone may be dominated by mangrove trees in tropical waters. This diversity of vertical structure creates foraging and nesting sites attractive to various bird species and functions as a segregating mechanism (Fig. 8.12). During breeding, some unique saltwater species are found there, such as the Clapper Rail and Seaside Sparrow, but also ubiquitous species like Red-winged Blackbird and Least Bittern. Wintering

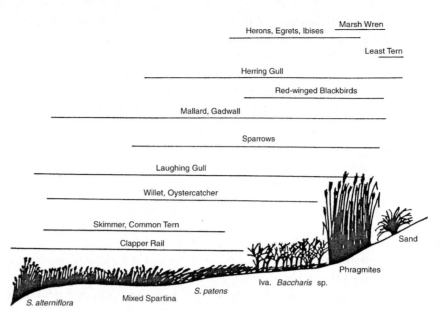

Herons, Egrets, Ibises   Marsh Wren

Least Tern

Herring Gull

Red-winged Blackbirds

Mallard, Gadwall

Sparrows

Laughing Gull

Willet, Oystercatcher

Skimmer, Common Tern

Clapper Rail

Sand

Phragmites

Iva. *Baccharis* sp.

S. patens

S. alterniflora   Mixed Spartina

Figure 8.12. Bisect of an Atlantic coast saltmarsh showing distribution and segregation of some common wetland birds in characteristic plant zonation (Burger (1985), with permission by Academic Press).

species include many freshwater, alkaline water, and wetland-edge birds from interior breeding areas (Sharp-tailed Sparrows, avocets, Willets, coots, and Northern Shovelers).

8.7.11 RIVERINE POOLS, SUBSTRATES AND SHORES

In their linearity and gradual increase from seasonal trickles of water to raging torrent and eventually to wide and permanent water at lower levels near their terminus in lake or sea, streams and rivers provide diverse shoreline configuration, vegetation, substrate and water characteristics, and water sources that are tapped by a great variety of birds. However, some stretches are nearly inhospitable at times because of the current while other stretches are characterized by flooding regimes that make them unpredictable. In higher areas, dippers and other passerines exploit adjacent vegetation and foods in and out of the water. Pools and backwater areas that hold fish and herps are favored by wait-feeders like Green Herons and night-herons, and pool and riffle areas are used by mergansers and other stream ducks mentioned above. Larger and slower streams have greater habitat diversity,which attracts a wider range of birds. River meanders are noted for moving basin materials of all sizes during high water periods and depositing them by size categories according to current and basin resistence. Thus, rivers cut against high ground and create vertical soil and sand banks suitable for kingfisher burrows, swallows, and other waterside birds. Depositional flats are formed on the inner courses, resulting in shallow water gradients suitable for waders and drier vegetated areas suitable for songbirds. Sandbars, islands and peninsulas are ideal for nesting by early succession species like Least Tern, Piping Plover, and Killdeer even when sparsely vegetated or bare. As these areas age, plant succession moves toward annual herbaceous vegetation and then to perennial plants, forming good areas for ground birds. Those areas grown up in shrubs, willows, and cottonwoods may form roosting areas for migrants, nest sites for colony-nesting egrets and herons, and perches for scavengers like eagles or piscivores like Ospreys.

8.7.12 BIRDS OF BOULDER, ROCK SUBSTRATE, AND OTHER HARDSHORES

Marine and large lakes can have large rocks and boulders deposited by forceful waves or flooded rock strata where aquatic organisms (plant or animal) occur; these areas attract birds that are specialists at probing and freeing attached foods. Constructed jetties, levees, and other stone or block structures simulate such natural sites and form substrates for invertebrates that attract birds. Many of the birds using rocky areas are strictly coastal, and they tend to be invertivores (especially limpet and mussel eaters), like Rock and Purple Sandpipers, Surfbirds, turnstones, oystercatchers, Harlequin Ducks, and eiders. Herbivorous birds that feed in these areas include Black Brant or their ecological counterpart in Tierra del Fuego, Kelp Geese, which use leafy sea-lettuce.

8.7.13 BIRDS OF CONSTRUCTED PONDS, LAKES, AND RESERVOIRS

Typically, constructed waters are impounded waters that are deep enough that emergent vegetation is either absent or restricted to a narrow fringe, and they sometimes have extensive submerged plant beds in the shore zone that are valuable for waterbirds. These artificial lakes may be important foraging sites for birds that feed on insects over the water, or for swimming birds that eat sub-mergent foliage or resident animals; as a result, they add avian diversity especially in arid regions (Evans and Kerbs 1977). Larger reservoirs may have extensive littoral zones in the upstream area, which may be excellent waterbird habitat if they are not subject to rapid and extreme water fluctuations. However, wave action often seriously limits shoreline plant establishment. Because of the abundance of fish, many are excellent areas for cormorants, pelicans, herons, and mergansers. Geese and similar grazers around the world use them as rearing areas in summer because they feed in the uplands and use the water for brood protection. Large reservoirs also may be favored by wintering flocks of Canada Geese and Mallards, because they are deep and remain open during most winters and the birds can feed in grain fields nearby (Simpson 1988).

Spatial and structural patterns

The habitat patterns described above are based on observed associations; although many have been quantified at study sites, fewer have been related to vegetation composition and structure by statisical tests. Patterns vary regionally, but they form a starting point for grouping microhabitats in many types of wetland and in wetland regions that should serve well in descriptive studies.

References

Bartonek, J. C. and Hickey, J. J. (1969). Food habits of canvasbacks, redheads, and lesser scaup in Manitoba. *Condor* 71, 280–90.

Beecher, W. J. (1942). *Nesting birds and the vegetative substrate.* Chicago, IL: Chicago Ornithological Society.

Bergman, R. D. (1973). Use of southern boreal lakes by postbreeding canvasbacks and redheads. *Journal of Wildlife Management* 37, 160–70.

Britton, J. C. and Morton, B. (1989). *Shore ecology of the Gulf of Mexico.* Austin, TX: University of Texas Press.

Brown, M. and Dinsmore, J. J. (1986). Implications of marsh size and isolation for marsh management. *Journal of Wildlife Management* 50, 392–7.

Brown, M. and Dinsmore, J. J. (1991). Area-dependent changes in bird densities in Iowa marshes. *Journal of the Iowa Academy of Sciences* 98, 124–6.

Buffington, J. M., Kilgo, J. C., Sargent, R. A., Miller, K. V., and Chapman, B. R. (1997). Comparison of breeding bird communities in bottomland forests of different successional stage. *Wilson Bulletin* 109, 314–19.

Burger, J. (1985). Habitat selection in temperate marsh-nesting birds. In *Habitat selection in birds,* ed. M. L. Cody, pp. 253–81. Orlando, FL: Academic Press.

Burger, J., Howe, M. A., Hahn, D. C., and Chase, J. (1977). Effects of tide cycles on habitat selection and habitat partitioning by migrating shorebirds. *Auk* 94, 743–58.

Burger, J., Shisler, J., and Lesser, F. H. (1985). Avian utilisation on six salt marshes in New Jersey. *Biological Conservation* 23, 187–212.

Capen, D. E. and Low, J. B. (1980). Management considerations for nongame birds in western wetlands. In *Workshop proceedings: Management of western forests and grasslands for nongame birds,* pp. 67–77. USDA Forest Service General Technical Report INT- 86. Washington DC: US Forestry Service.

Celada, C. and Bogliani, G. (1993). Breeding bird communities in fragmented wetlands. *Bolettino Zoologia* 60, 73–80.

Colwell, M. A. and Dodd, S. L. (1997). Environmental and habitat correlates of pasture feeding by nonbreeding shorebirds. *Condor* 99, 337–44.

Cowardin, L. M., Carter, V., Golet, F. C., and LaRoe, E. T. (1979). *Classification of wetlands and deepwater habitats of the United States.* FWS/OBS-79/31. Washington, DC: US Fish and Wildlife Service.

Craig, R. J. and Beal. K. G. (1992). The influence of habitat variables on marsh bird communities of the Connecticut River estuary. *Wilson Bulletin* 104, 295–311.

Daiber, F. C. (1982). *Animals of the tidal marsh.* New York: Van Nostrand Reinhold.

Damman, A. W. H. and T. W. French. (1987). *The ecology of peat bogs of the glaciated*

Northeastern United States: a community profile. Report 85(7). Washington, DC: US Fish and Wildlife Service.

DeGraaf, R. M. (1986). Urban bird habitat relationships: application to landscape design. *Transaction of the North American Wildlife and Natural Resource Conference* 51, 232–48.

Derksen, D. V., Eldridge, W. D., and Weller, M. W. (1982). Habitat ecology of Pacific Black Brant and other geese moulting near Teshekpuk Lake, Alaska. *Wildfowl* 33, 39–50.

Duebbert, H. F. (1982). Nesting of waterfowl on islands in Lake Audubon, North Dakota. *Wildlife Society Bulletin* 10, 232–7.

Esler, D. (1992). Habitat use by piscivorous birds on a power plant cooling reservoir. *Journal of Field Ornithology* 63, 241–392.

Evans, K. E. and Kerbs, R. R. (1977). Avian use of livestock watering ponds in western South Dakota. General Techical Report RM-35. Washington, DC: US Forest Service.

Faaborg, J. (1976). Habitat selection and territorial behavior of the small grebes of North Dakota. *Wilson Bulletin* 88, 390–99.

Flake, L. D. (1979). Wetland diversity and waterfowl. In *Wetland functions and values: the state of our understanding,* eds. P. E. Greeson,, J. R. Clark, and J. E. Clark, pp. 312–19. Minneapolis, MN: American Water Resources Association.

Gibbs, J. P., Longcore, J. R., McAuley, D. G., and Ringelman, J. K. (1991). *Use of wetland habitats by selected nongame water birds in Maine.* Fish & Wildlife Research No. 9. Washington, DC: US Fish & Wildlife Service.

Glazener, W. C. (1946). Food habits of wild geese on the Gulf Coast of Texas. *Journal of Wildlife Management* 10, 322–9.

Harrington, B. (1996). *The flight of the Red Knot.* New York, NY: W. W. Norton.

He, F. and Legendre, P. (1996). On species–area relations. *American Naturalist* 148, 719–37.

Hehnke, M. and Stone C. P. (1978). Value of riparian vegetation to avian populations along the Sacramento River system. In *Strategies for protection and management of floodplain wetlands and other riparian ecosystems,* Technical Coordinators R. R. Johnson and J. F. McCormick, pp. 228–35. General Technical Report WO-12. Washington, DC: US Forest Service.

Hirano, T. and Higuchi, H. (1988). The relationship between river width and the occurrence of riparian bird species in winter (with English summary). *Strix* 7, 203–12.

Hobaugh, W. C. (1984). Habitat use by snow geese wintering in southeast Texas. *Journal of Wildlife Management* 10, 1085–96.

Hodges, M. F. Jr and Krementz, D. G. (1996). Neotropical migratory breeding bird communities in riparian forests of different widths along the Altamara River, Georgia. *Wilson Bulletin* 108, 496–506.

Kaminski, R. M. and Prince, H. H. (1984). Dabbling duck–habitat associations during spring in Delta Marsh, Manitoba. *Journal of Wildlife Management* 10, 37–50.

Kaminski, R. M. and Weller, M. W. (1992). Breeding habitats of Nearctic waterfowl. In *Ecology and management of breeding waterfowl,* eds. B. D. J. Batt, A. D. Afton, M. G. Anderson, C. D. Ankney, D. H. Johnson, J. A. Kadlec, & G. L. Krapu, pp. 568–89. Minneapolis, MN: University of Minnesota Press.

Keller, C. M., Robbins, C. S., and Hatfield, J. S. (1993). Avian communities in riparian forests of different widths in Maryland and Delaware. *Wetlands* 13, 137–44.

Lampolahti, J. and Nuotio, K. (1993). Umpeenkasvu koyhdyttaa lintuvesia (with English summary). *Linnut* 28, 13–17.

MacArthur, R. H. and Wilson, E. O. (1967). *The theory of island biogeography*. Princeton, NJ: Princeton University Press.

McCabe, R. A. (1991). *The little green bird: ecology of the willow flycatcher*. Madison, WI: Rusty Rock Press.

Murkin, H. R., Kaminski, R. M., and Titman, R. D. (1982). Responses by dabbling ducks and aquatic invertebrates to an experimentally manipulated cattail marsh. *Canadian Journal of Zoology* 60, 2324–32.

Murkin, E. J., Murkin, H. R., and Titman, R. D. (1992). Nectonic invertebrate abundance and distribution at the emergent vegetation-open water interface in the Delta Marsh, Manitoba, Canada. *Wetlands* 12, 45–52.

Nilsson, L. (1978). Breeding waterfowl in eutrophicated lakes in south Sweden. *Wildfowl* 29, 101–10.

Nudds, T. D. (1992). Patterns in breeding waterfowl communities. In *Ecology and management of breeding waterfowl*, eds. B. D. J. Batt, A. D. Afton, M. G. Anderson, C. D. Ankney, D. H. Johnson, J. A. Kadlec, & G. L. Krapu, pp. 540–67. Minneapolis, MN: University of Minnesota Press.

Orians, G. H. (1980). *Some adaptations of marsh-nesting blackbirds*. Princeton, NJ: Princeton University Press.

Patterson, J. H. (1976). The role of environmental heterogeneity in the regulation of duck populations. *Journal of Wildlife Management* 40, 22–32.

Simpson, S. G. (1988). Use of the Missouri River in South Dakota by Canada Geese in fall and winter, 1953–1984. In *Waterfowl in winter*, ed. M. W. Weller, pp. 529–40. Minneapolis, MN: University of Minnesota Press.

Skagen, S. K. and Knopf, F. L. (1994). Migrating shorebirds and habitat dynamics at a prairie wetland complex. *Wilson Bulletin* 106, 91–105.

Spence, D. H. N. (1982). The zonation of plants in freshwater lakes. *Advances in Ecological Research* 12, 37–125.

Swanson, G. A., and Meyer, M. L. (1977). Impact of fluctuating water levels on feeding ecology of breeding blue-winged teal. *Journal of Wildlife Management* 41, 426–33.

Swift, B. L., Larson, J. S., and DeGraaf, R. M. (1984). Relationship of breeding bird density and diversity to habitat variables in forested wetlands. *Wilson Bulletin* 96, 48–59.

Taylor, D. M. (1986). Effects of cattle grazing on passerine birds nesting in riparian habitat. *Journal of Range Management* 39, 254–8.

Tuck, L. M. (1972). *The Snipes: a study of the genus Capella*. Monograph Series No. 5. Otttawa: Canadian Wildlife Service.

Tyler, S. J. and Ormerod, S. J. (1994). *The dippers*. London: T & A D Poyser.

Wauer, R. H. (1977). Significance of Rio Grande riparian systems upon the avifauna. In *Importance, preservation, and management of riparian habitat: a symposium*, Technical Coordinators R. R. Johnson & D. A. Jones, pp. 165–74. General Technical Report RM-43. Washington, DC: US Forest Service.

Weller, M. W. (1979). Birds of some Iowa wetlands in relation to concepts of faunal preservation. *Proceeding of the Iowa Academy of Science* 86, 81–8.

Weller, M. W. (1988). Bird use of an east Texas shrub wetland. *Wetlands* 8, 145–58.

Weller, M. W. (1996). Birds of rangeland wetlands. In *Rangeland wildlife*, ed. P. R. Krausman, pp.71–83. Denver, CO: Society for Range Management.

Weller, M. W. and Fredrickson, L. H. (1974). Avian ecology of a managed glacial marsh. *Living Bird* 12, 269–91.

Weller, M. W. and Spatcher, C. E. (1965). *Role of habitat in the distribution and abundance of marsh birds*. Special Report No. 43, pp. 1–31. Ames, IA: Iowa State Agriculture and Home Economics Experiment Station.

Whitlack, R. B. (1982). *The ecology of New England tidal flats: a community profile*. OBS-81/01. Washington, DC: U S Fish & Wildlife Service.

Withers, K. and Chapman, B. R. (1993). Seasonal abundance and habitat use of shorebirds on an Oso Bay mudflat, Corpus Christi, Texas. *Journal of Field Ornithology* 64, 382–92.

Woodin, M. C. and Swanson, G. A. (1989). Foods and dietary strategies of prairie-nesting Ruddy Ducks and Redheads. *Condor* 91, 280–7.

Zimmerman, J. L. (1990). *Cheyenne Bottoms, wetland in jeopardy*. Lawrence, KS: University Press of Kansas.

9

Habitat dynamics:
water, plant succession, and time

9.1 Water variability

The dynamics of water over time, whether seasonal, annual, or longer term, dictates the chemical and physical character of wetland water, the resulting vegetation, and the use of wetlands by birds and other aquatic or semiaquatic life. Numerous authors have summarized the consequences of variable water regimes on vegetation, plant succession, and size and depth of various types of wetlands (Bellrose, Paveglio and Steffeck 1979, Chabreck 1988, Golet and Parkhurst 1981, Gosselink 1984, Kantrud, Krapu and Swanson 1989, Kushlan 1989, Stanley, Fisher and Grimm 1997, Weller and Fredrickson 1974). Here we examine the general patterns of water influences on vegetation and the physical aspects of the wetland habitat that influence bird use. Examples also are given that show the similarities of these patterns worldwide.

9.2 Temporal changes in wetland vegetation

Because of the importance of vegetation structure and food resources in attracting a diversity of birds, we must focus on some of the factors that influence (i) the establishment of the plants in wetlands, based on their life-history strategy; (ii) the influence of water depths and substrates on plant survival and reproduction; and (iii) the dynamics of plant species and vegetation patterns (**plant succession** or **biotic change**). Birds respond to many of the same environmental influences as do the plants and, thus, are intimately linked with the entire succession process, regardless of the time frame.

Water is the major influence on wetland habitat variability, often changing seasonally and from year to year. Because many wetland plants do not germinate or survive in deep water, declining water levels may provide opportunities for seeding events, but stability of water levels at key times to ensure survival and growth thereafter obviously influences the species composition and health of the plant community (Fredrickson & Reid 1990). Coupled with seasonal temperature changes, water is the major influence on chronology and species composition of seasonal vegetation development. Long-term changes in the plant community also may be caused by changes in nutrient availability of the growth medium through plant biomass production, deposition, and siltation trapping, which provide opportunities for other plant species to establish while earlier pioneers decline (Carpenter 1981). Competition among plants also may be important (Bertness and Ellison 1987). In some communities, fire (Gunderson, Light and Holling 1995) and grazing (Kruse and Bowen 1996, Reimold, Linthurst and Wolf 1975) also influence changes in plant species composition and the rates at which changes occur. Moreover, such short-term and often cataclysmic changes can have long-term implications, especially with long-lived plant species.

9.3 Plant succession

A wetland plant community may be established or re-established when suitable water conditions and biochemical properties of the soil induce germination of seeds or growth of plant propagules (e.g., tubers and rhizomes) of hydrophytes (van der Valk 1981). The viability, mobility, and response of seeds to suitable germination conditions have been well documented (Crocker 1938, van der Valk and Davis 1978). This is perhaps best demonstrated in small wetland basins or along fringing shorelines of larger wetlands where gradually deepening water depth determines which species germinate, spread, and survive according to their individual adaptations for particular water and soil conditions. As discussed earlier, the structure or life form of various species tends to vary with water depth, but this is strongly influenced by the taxonomic groups dominant in the locale. Such plant zones also are clear-cut in the submerged vegetation of shallow lakes, where it influences accretion rates and increases lake eutrophication (Carpenter 1981), and in shallow wetlands, where competition as well as physiological responses to conditions at those sites may influence plant species competition and success (van der Valk and Welling 1988, Wilson and Keddy 1985).

Coastal estuarine and salt marsh wetland habitats have similar patterns of zonation, which are induced and influenced by daily and seasonal variability in tidal regimes (Allison 1992, Niering and Warren 1980, Penfound and Hathaway 1938). Drying in higher zones increases salt concentration, which may reduce plant growth or influence which species can survive; however, in lower tidal areas, freshwater inflows may dramatically change plant survival and species composition (Allison 1992). Wave action influences plant success in deeper zones, and as in all plant communities, competition may influence the plant species that are present and their success in addition to the effects of water regimes (Pennings and Callaway 1992).

The resulting vegetation structure along shorelines is an important one for foraging and nesting birds, and for those seeking shelter and protective cover. The vegetation can also be important because of differences in the potential foods produced, e.g., annuals tend to produce more seeds and perennial species tend to produce more tubers or rhizomes (Chapter 5). As a result, shoreline gradients that induce zonation of structure and food resources provide diverse resources to which birds respond dramatically (Allen 1914, Beecher 1942, Weller 1994, Weller and Spatcher 1965), resulting in segregation through habitat selection by various bird species. But keep in mind that what is an attractive structure for some bird species will discourage others, for example those that favor bare mudflats (shorebirds) or short, flooded vegetation (yellowlegs and certain teal). Presumably, these structural patterns are signals to bird species about the habitat resources.

The distribution of vegetation (i.e., pattern) in more central areas of a wetland are especially relevant to swimming birds and others that favor more open water, like the large waders. The development of such patterns are complex because they represent historical seeding or other short-term plant establishment events that themselves reflect various rates of survival dependent on water regimes, and wave and ice action.

Clearly, patterns of plant distribution are not fixed but change over time based on the wetland type, water regimes, and plant life-history strategies. To appreciate the role of such changes for birds, conceptual models of biotic change have been useful (Begon, Harper and Townsend 1996), and two have been especially relevant to wetlands (van der Valk and Davis 1978). These differ mainly in temporal scale but also in their assessment of the variability of individual sites and the influence of chance events on the long-term history of vegetation at a site.

The traditional view of long-term changes is based mainly on work by Clements (and, therefore, termed Clementsian succession) and suggests that water-driven plant (hydrarch) succession proceeds from wet to dry in a series of stages (called "seres"). It is considered very long-term (hundreds to thousands of years), directional, and fairly predictable. Examples of long-term succession in northern bogs (where water is more stable) have been based on analysis of pollen or planktonic fossils, which show changes over tens of thousands of years. At this scale, models of changes in bird species over time have been based on the synthesis of observations on concurrently observed examples of various seres (Aldrich 1945). But even these habitat areas are not stable and often experience reversals reflecting long-term climate change or other variables like salinity and sedimentation. Lake succession has been viewed in this long-term scale, reflecting eutrophication via sedimentation and plant organic deposits within the basin; submergent aquatic plants are especially important in sediment trapping (Carpenter 1981, Robel 1961). These plants influence use by aquatic birds and even serve as indicators of these lake stages (Kauppinen and Vaisanen 1993). Ultimately, lakes may fill in with sediments and bog plant growth, creating sites for pioneering shrubs and trees (Fig. 9.1).

Other plant ecologists, especially Gleason (hence, Gleasonian succession), felt that succession was much more influenced by specific site conditions and

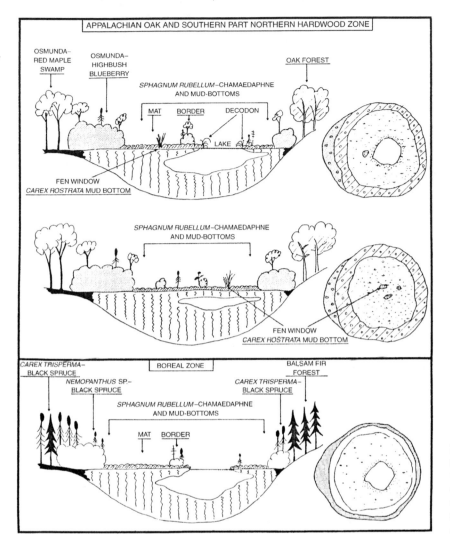

Figure 9.1. Long-term successional changes in forested bog wetland habitats in the northeastern USA, showing reduced open water and increased shrub and tree invasion (Damman and French 1987).

by chance environmental events, which result in differing patterns of vegetation establishment even within the same area or site (van der Valk 1981). Wetland basins that dry periodically present random opportunities for different seeds to germinate, and hence can, by chance, result in strikingly different vegetation in the same basin over a series of years. A series of observations by different workers studying basin wetlands demonstrated dramatic short-term changes in plant succession induced by **drawdowns** (also termed drydowns in some areas) caused by drought (or water-level manipulations for management purposes); these resulted in ideal germination conditions from the buried seed bank or residual rhizomes (Harris and Marshall 1963) (Fig. 9.2). Animal responses to these patterns also were noted over several years (Kadlec 1962, Meeks 1969, Weller and Spatcher 1965). But there are also records

Habitat dynamics

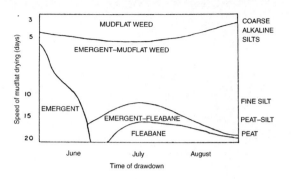

Figure 9.2. Model of the chronology of seed germination in response to water drawdown in a shallow northern Minnesota wetland (Harris and Marshall 1963, with permission).

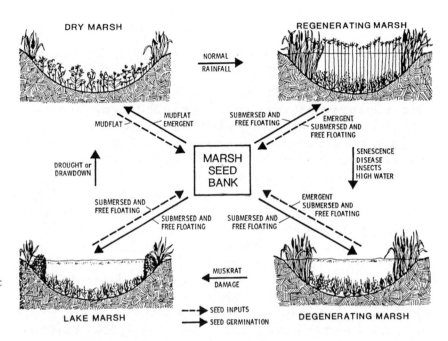

Figure 9.3. Model of marsh revegetation and survival of plants as influenced by dynamic water levels of a freshwater wetland (van der Valk and Davis 1978, with permission).

of this same type of semipermanent wetland that seems to have had the same vegetation for thousands of years. Many invertebrates also have residual and drought-tolerant propagules or egg banks that allow a quick response and recolonization when suitable water conditions occur. Others must fly in and lay eggs (dragonflies), as is true of birds.

Increases or decreases in water depth, changes in hydroperiod (the duration of flooding), or water characteristics such as turbidity influence the success of plants established during low water. These form a somewhat predictable pattern that varies with water, timing, and herbivory (Fig. 9.3).

The plant reproductive strategies discussed earlier are changed in these ways and some species decline while others increase over several growing seasons. Submergent plants are very dependent on water depth and clarity, and modest changes can affect their success. In both fresh- and saltwater wet-

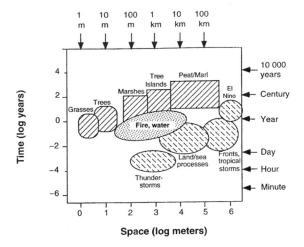

Figure 9.4. Role of time, water, and fire in the development and maintenance of vegetation types in the Florida Everglades. (Reproduced with permission from Gunderson *et al.* (1995) *Bioscience* (Supplement) 566–573. © American Institute of Biological Sciences.)

lands, dry periods drastically curtail submergents (commonly perennial), while marginal annual emergent plants may flourish. But excessive water can deter even highly aquatic emergent plants from reproducing vegetatively. In taller vegetation such as shrub or forested wetlands, changes in the amount of sunlight reaching the substrate or ground-level plants owing to die-off or excess growth can induce major changes in understory plants. Where plant communities are lost for any reason, re-establishment often is slow if conditions inducing germination of the seed bank are not present (i.e., drawdown phase for germination), and open-water wetlands or bays may result.

Temporal scales and influential factors that affect succession are nicely shown in a model of the development of vegetation in the Florida Everglades and the role of fire (Fig. 9.4) (Gunderson *et al.* 1995). Organisms at all trophic levels respond to these plant changes, because they may be used directly as food but also because the major detritivores in a wetland system, the invertebrates, are quick to reproduce and bloom on these new nutrient bases. The invertebrates form major food resources for waterbirds and have been much studied in relation to natural and experimental successional patterns (Murkin and Kadlec 1986, Murkin, Kaminski and Titman 1982, Voigts 1976) (Fig. 9.5).

9.4 Bird responses to short-term fluctuations in water regimes

Several examples of bird use of wetland types by water regime have been given. To provide a framework for understanding these changes, I would like here to summarize the general patterns noted by many authors, and discussion below will enlarge on these behaviors as they relate to bird conservation and management.

Wetlands with variable water levels probably are populated with invertebrates (Voigts 1976, Wiggins, MacKay and Smith 1980) and plants (Weller and

Habitat dynamics

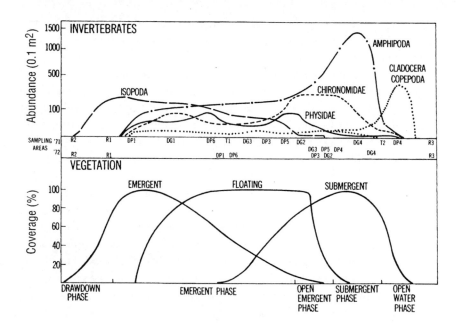

Figure 9.5. Model of the abundance of invertebrate dominants in relation to dynamics of vegetation based on pooled data from several basin wetlands (Voigts 1976, with permission).

Spatcher 1965) that are best adapted to such conditions. Responses by birds differ with the stage of the life cycle, the birds being most sensitive to change during the nesting period. Relatively rapid changes in water levels have a direct impact on nesting birds by flooding out nests or by drying protective water moats so that nests are exposed to predation. Abandonment varies by species, and colonial nesters are more likely to abandon en masse (e.g., terns and some gulls), although the important clues are unknown. Individual variation is common among ducks and coots that nest over water, and this probably also varies by region. For example, Redhead ducks in semipermanent marshes abandon nests at the slightest drying at the base of the nest, whereas they may be on totally dry ground in desert areas in the western USA. Passerines seem to abandon if the drying occurs early but are more attached to the nest if the young are hatched. Species with mobile broods like ducks and rails simply move overland to wetter areas.

It is in foraging that we see the most rapid and dramatic response to water-depths changes. Regardless of the scale of mobility, species seem always to be searching for sites with higher densities of food and increased vulnerability for higher rates of consumption. If those sites are not found within a wetland, they move elsewhere. Species that favor more permanent and relatively stable water include swimming birds that feed on fish or large invertebrates such as grebes, loons, cormorants, pelicans, and sea ducks. Slightly rising or declining waters do not deter most of these, but obviously food is more important that water depth as long as a level of location and safety from predators is reached.

Birds that respond quickly to declines in water level include many shorebirds (e.g., stilts, sandpipers, and plovers), dabbling ducks, egrets, herons, storks, ibises, and shoreline passerines. All are tapping foods that are available, but each group favors different foods suitable to diverse feeding tactics. Shorebirds and passerines may be competing for crustaceans and insects exposed by or attracted to the mudflat or vegetation. Teal may be straining on the mudflats with the probing or pecking shorebirds but taking still smaller prey. The walking fish feeders like egrets, ibises, and storks (and sometimes large Icterids like grackles) are exploiting fish or herptiles concentrated in drying pools. Birds that favor feeding in vegetation tend to be reduced (rails, bitterns, and some sparrows and ducks).

In situations when wetlands dry completely, vegetation invasion influences use, and wetland birds are more likely to be replaced by species that prefer damp sites (plovers) and upland shorelines (doves, passerines). When water levels rise beyond the basin shoreline, flooding shoreline vegetation and adjacent meadows and open fields, many species find a diversity of foods available, which they may pursue as flocks even though they are more typically solitary birds. These include some some larger shorebirds like stilts and curlews, ibises, storks, dabbling ducks, egrets, and small herons.

9.5 Seasonality of bird responses to water and vegetation

Seasonality is most dramatic and regular at high latitudes (northern or southern hemispheres) as a result of temperature changes as well as precipitation. Seasonal changes in tropical and warm-temperate areas often result from rainfall patterns. The effects of these on vegetation establishment, growth, and survival were described above, and their potential influence on birds is obvious. Because birds are highly mobile and migratory at high latitudes, and nomadic in drier and warmer areas, we must always keep in mind that such regional water dynamics have the potential of influencing the direction and timing of bird movements and colonization.

To examine possible seasonal affects of water regime on waterbirds and the responses of vegetation, we will examine water patterns by wetland type. Figure 9.6 presents some hypothetical but observation-based diagrams of water regime changes by selected wetland type by season. As discussed earlier, a lake is a lake because of the greater stability of deeper water. Therefore, we do not expect as much water-level fluctuation in a lake-like (lacustrine) wetland or a constructed reservoir as we do in a more shallow, seasonally flooded basin. This results in greater opportunity for the establishment of annual or perennial hydrophytes through germination or rhizome spread. Whether a constructed lake (reservoir) varies depends on water uses; some irrigation reservoirs change daily or weekly, while flood control reservoirs release water in advance of flood so they can store high runoff during the rainy season. Obviously, this water fluctuation influences use by birds in the reservoir, and

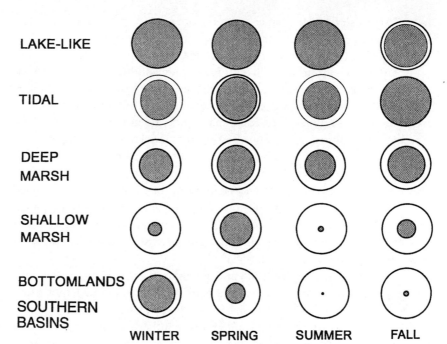

Figure 9.6. Generalized diagram of common seasonal water patterns of various wetland types and areas in North America as they influence water availability and vegetation for migratory birds.

by fish downstream; it can be catastrophic for nesting birds or attractive to socially feeding waders exploiting trapped fish.

Tidal regimes are characterized by mean values reflecting the high and low for the tide cycle; these vary greatly in different parts of the world because of latitude and continental configurations, which influence water movement. But tides also vary by season and they vary from year to year because of astronomical relationships of the moon and sun. Such cycles may last several years, and at least in our experience with the modest tides (12–16 inches (30–40 cm)) of the Texas Gulf of Mexico, they can cause water-level changes sufficient to influence vegetation and, therefore, use by birds over a period of several years (Weller 1994); this pattern is comparable to that seen in freshwater basin wetlands (Weller and Fredrickson 1974).

Birds that migrate annually to shallow wetlands to breed or feed en route to more northerly areas could encounter great seasonal and year-to-year variation. Birds that are dependent on perennial vegetation like cattail and bulrushes for nesting locally would seek out wetlands with more permanent waters. Thus, the typical assemblage of birds of shallow wetlands are those that use low vegetation, seeds of annual plants, and wet-soil invertebrates and move readily to adjacent wetlands as needed: yellowlegs, phalaropes, Sedge Wren, several rails, perhaps Red-winged Blackbirds, and possibly Blue-winged Teal if pools occur. Birds of more permanent marsh include those dependent upon robust vegetation, like Yellow-headed Blackbirds, bitterns, and herons, and swimmers like coots, grebes, and ducks, all of which tend to use invertebrates, herptiles, or fish typical of larger and more open pools of water. Large waters like lakes and reservoirs may have few birds, and those will be ones dependent upon large fish (gulls, loons, cormorants, and Ospreys).

In any given season, birds will respond to water conditions initially, probably more than to vegetation; residual plants are more important to perching birds establishing a breeding territory than to swimming birds searching for food and for loafing sites. If typical water regimes do not exist because of the extremes of drought or heavy rains, nesting may be delayed (Custer, Hines and Custer 1996, Weller and Spatcher 1965), may not occur (Breeden and Breeden 1982, Frith 1967, Rogers 1959), or some species may move to where conditions are more suitable to their preferences (supported by the re-capture of banded ducks and observations of conspicuous colonial species like storks and egrets). Therefore, populations may decline (Weller, Wingfield and Low 1958) and the species composition of nesting birds may change to those characteristic of the new water regime (Weller 1994).

Why do few wintering birds stay and nest in southern wetlands rather than engage in hazardous migration twice a year? If we examine the hypothetical pattern of deep marsh and southern basins or bottomland wetlands (Fig. 9.6), it is obvious that birds that do not move from southern to northern breeding marshes may face drying wetlands too early to rear young. A few species associated with coastal tidal waters (e.g., Black-necked Stilts and Willets) are more likely to be successful. These typical patterns can be changed drastically by heavy late summer or autumn rains resulting from hurricane or tropical storms, and sufficient water may carry over to keep wetlands flooded for several years. In that situation, many species may remain and breed, although the numbers probably are modest compared with the total migrant population. Even then, most birds are more likely to find the essential requirements for breeding by moving from areas of few wetlands to one of more numerous and diverse wetlands, where some wetlands within a complex of various-sized basins are likely to be wet.

In autumn, southern basins may be near dry and unsuitable for autumn migrants – especially those birds that require specific water conditions such as mudflats for shorebirds or flooded areas for birds that feed by swimming. Migrants then move through quickly on northern storm fronts, and little wintering would be expected. If flooded by autumn rains, these areas are suitable for late migrants and presumably are more heavily used in those wet winters.

Seasonality of nest initiation seems to be influenced by pre-migration and pre-nesting resources but are modified by temperature and other stressors. Winter body condition seems to influence timing of migration, readiness for pairing and laying eggs, and survival during this more stressful period (Heitmeyer and Fredrickson 1981). These activities are mostly a product of wetland habitat conditions, including presence and timing of open water (thawing at high latitudes) and availability of suitable foods and nest sites. For any given region, data are available that show the general chronology (i.e., timing) of nest initiation, and this presumably reflects those long-term patterns of pre-migratory preparation, migration, finding suitable nesting cover/sites, and pair formation, where that is done on breeding areas. But if the phenology (climatically influenced events) differs, and the season is "late" (i.e., cold and wet) or "early," nesting often is delayed or advanced accordingly in many species. Habitat suitability clearly plays a major role in regulating these events.

9.6 Bird responses to long-term changes in water and vegetation

9.6.1 SMALL BASIN WETLANDS IN THE PRAIRIE POTHOLE REGION OF NORTH AMERICA

When drought strikes the agricultural farmlands of the Upper Midwest in the USA, everyone is made aware of the impact on food production, costs to consumers, and hardships to landowners. The famous drought of the 1930s is best known because of other concurrent economic stresses, but similar droughts have occurred several times in recent history. Still more recently, the great flood of the Missouri and Mississippi River system in 1993 demonstrated how the loss of upland catch-basins had a major impact on river flood plains (Miller and Nudds 1996). Moreover, these rains filled lakes and small basin wetlands to the highest levels ever recorded in the northern USA and southern Canadian prairies. But how did birds respond to these extremes?

Over the years of annual waterfowl aerial surveys and banding, fair documentation has provided insights into at least duck responses, but the record is not simple and as well documented as we would like because the scale is so large. It is known, for example, that local populations of many waterbirds respond quickly up or down in individual wetlands, but regional impacts on populations and species richness probably are softened by the availability of wetland complexes of different sizes, which allows birds to shift from one wetland to another and still successfully reproduce in the area. In major droughts, however, there is good evidence that continent-level shifts occur. Best documented is the strong-flying pintail, which prefers low vegetation and open areas and shifts into the Arctic (including Alaska and Siberia) when the prairie potholes go dry. How successful they are is uncertain, but they seemingly survive better than they might sitting around drying wetlands where they would be vulnerable to predators and where nesting success would be unlikely. There are observations that suggest that other species may shift to marginal areas to the west and south of the central duck-breeding area, if the areas happen to be wet. Data on other birds are less well documented, but we have every reason to believe that they too are highly mobile and make major spatial shifts. Recent responses of duck populations to wet cycles in the Prairie

Pothole Region are impressive, with the highest populations recorded in many years (Caithamer and Dubovsky 1997).

9.6.2 SOUTH TEXAS COASTAL PONDS

South Texas is known as a dry place even along the coast. The long-term mean rainfall is, however, about 70cm (27 inches) per year, coming mainly in the periodic deluges that result from hurricanes and similar tropical summer storms. Between these flood years, it is very dry. Even most knowledgeable naturalists would be surprised to find that there are numerous wetlands there, sometimes termed "hurricane lakes," because they are not there all the time. Even during relatively dry periods, they are favored grazing areas for livestock and deer because they catch and hold whatever rain falls and stimulates winter plant growth. But when the rains come, 5 or even more inches (13cm) fall in 24 hours, and the rains will come repeatedly over a few weeks or months, the area rivals the Prairie Pothole Region of the northern USA and Canada. Terrestrial grasses are no more. Sedges and other moist-soil plants seem to spring up, and even a small Blue Water-lily seems to explode in shallow, freshwater wetlands after drought and re-flooding regardless of whether grazing has occurred in previous years. Wintering waterfowl increase in the area, driven by the need for sources of fresh water after feeding on saline seagrasses. Ducks such as Blue-winged Teal, Northern Shovelers, and Ruddy Ducks and American Coots, Moorhens, and rails settle in and nest. Some wetland mapping systems may not regard these as wetlands because of the long interval between hydric plant communities, but clearly they have the residual seed banks to respond when water is present.

9.6.3 LARGER FRESHWATER BASIN WETLANDS

Several of our studies of wetlands (Weller and Fredrickson 1974, Weller and Spatcher 1965) of 100 to 400 acres (40 to 162 ha) in central and northwest Iowa demonstrated the impact of vegetation change on species composition and abundance of wetland bird assemblages. Species richness was greatest when a 50:50 up to 30:70 ratio of cover:water existed, as was abundance of most birds that favored robust emergents. Species such as Black Terns and Red-winged Blackbirds that are attracted to shallow and dense marsh vegetation lost out to Forster's Terns and Yellow-headed Blackbirds under the changes to deeper water and more robust emergent plants.

In still larger basins of thousands of acres, I had an opportunity to spend some time observing vegetation and birds of the managed wetland called Mud Lake on the Agassiz National Wildlife Refuge in northern Minnesota, where Harris and Marshall (1963) developed one of the earliest models of seed-based succession. They recommend that drawdowns be done periodically as needed to maintain the variety and dispersions of emergents needed to attract nesting waterfowl to the area. However, because of concerns for botulism (a disease that may flourish during drawdown conditions and kill thousands of ducks), and because Mud Lake was providing rich beds of submergent foods for migrant waterfowl, one pool was not drawn down and more and more open water resulted. But nearly 25 years after their study, the wetland had to be drained for repair of the water-control structure, providing a unique opportunity to observe the resultant re-growth. In addition to verifying the germination patterns observed earlier, we observed mounds of cattail growing on layers of rootstocks that had built up with the deep water. When the lake was dewatered, these clumps of cattail still survived and the mounds of rootstocks stood up in the mudflats about 2 ft (60 cm) high. When a few inches of water returned to the flats to stimulate growth of shallow-marsh species like spike-rush and Soft-stemmed Bulrush, these islands of cattail formed vertical walls of cattail roots nearly 2 ft above the water level. If not flooded, these areas would have returned to the marginal willow-carr, a combination of sedge meadow interspersed with shrub willow, which is common to these northern wetlands that are only seasonally flooded. The bird life that resulted also differed strikingly from the open-water period, as Pied-billed Grebes, American Coots, diving ducks, and Yellow-headed Blackbirds moved in and nested abundantly in the newly established vegetation where nest sites were sparse before. Despite the shallow water, submergents re-established as well and provided food for young ducks in late summer and for migrant ducks in the fall.

9.6.4 SOME ARGENTINE WETLAND DYNAMICS

In Argentina, once vast marshes have now been drained and modified, but they are still significant in the Pampas Region and are comparable to the grasslands of North America and Australia. Many of these marshes were semi-permanent and large because of the flat topography. In 1964, I went to Argentina in search of the Black-headed Duck, a completely parasitic species that neither builds a nest nor rears its own young. There I found significant numbers in several marshes, where they laid their eggs in the nests of coots and other marsh birds, but I was told (and the vegetation supported this conclusion) that the marshes had been dry and that conditions were not at their best. I next returned to these marshes 10 years later, when my local naturalists advised that the area had been dry since my departure and the species had been seen only rarely. During that visit, however, the marshes were flooded and Black-headed Ducks were widespread. Two Canadian biologists recently have worked in the same marshes and have within 3 years seen the same kinds of fluctuation I saw over 10 years. Obviously, these birds must be as mobile their hosts to survive such wetland fluctuations. Moreover, population size must be highly variable.

Australia is better known for its extensive deserts than for its wetlands, but both may occur side by side in response to torrential rainfall and flooding. Waterbird breeding habitat can be highly variable in time and space (Frith 1967), producing the same boom-or-bust waterbird population pattern we see in the North American Prairie Pothole Region. However, waterfowl biologists there have found less seasonality in breeding than is typical of northern hemisphere wetlands; they do see an immediate response to water by many species but with an unexpected variance in response time by different species (Frith 1967, Briggs 1992). When examining breeding chronology of nearly 50 species of waterbirds, most did breed in spring but few seemed to respond to photoperiod changes, and most initiated breeding on rainfall clues (Halse and Jaensch 1989). Moreover, ducks, like passerines, initiate long flights in response to rainfall farther away than most thought they could detect. Then, apparently through responses to feeding adaptations and food availability, various duck species started breeding at various times from flood to drydown stages. Discovery of these patterns in Australia led North American biologists to recognize similar patterns in North America (Rogers 1959). Waterbirds in Australia also respond to flooding patterns much as seen elsewhere: re-flooded dry basins are more attractive to breeding birds than are simple increases in the levels of already flooded basins, presumably because of rich and diverse invertebrate and seed resources for filter feeders. More predaceous or piscivorous birds like egrets and cormorants favored more permanent water with fish populations (Crome 1988).

9.6.6 THE FLORIDA EVERGLADES

The Florida Everglades is one of the largest wetlands in North America, and one of the most stressed in terms of human modification by drainage and water-level regulation. This subtropical wet savannah undergoes seasonal and long-term variations in water regimes that are responsible for its vegetation and its avian uniqueness. The rain cycle in the Everglades area is one of early summer and winter drying, with the water levels varying from year to year and seasonally dependent upon the amount of rain. When vast areas dry, they leave pools containing fish and amphibians that are the food for Wood Storks, ibises, and egrets and herons (Browder 1978, Kushlan 1989). But there also are deeper pools of emergent vegetation in the Everglades that support species like Apple Snails, which require permanent water. When even the deeper areas go dry, and they are dry for too many years, they no longer support the Apple Snails on which both endangered Snail Kites and Limpkins feed.

When the Everglades was but one of many large wetland areas in Florida (the Kissimmee and St John's river systems now are being restored), it is probable that movement from one area to another helped to maintain waterbird populations. Moreover, current-day levees and impoundments have modified the wetting and drying patterns so that pool distribution and water regulation is no longer natural. The impact on waterbirds has been great and has produced serious controversy among conservationists because of a lack of understanding of waterbird mobility and population expectations in a variable environment. Unfortunately, the two most endangered birds, the Snail Kite and the Wood Stork, require just the opposite conditions. By virtue of the US Endangered Species Act, federal agencies must maintain the highest populations of both whereever they occur. However, these birds probably always had variable populations because they moved for feeding and breeding to those places that met their needs – as seems characteristic of wetland birds all over the world. Wood Storks are strong fliers and seek out drying areas where fish are vulnerable; therefore, they have shifted feeding areas regionally to avoid flooded management pools and have even shifted breeding areas farther north into Georgia. Snail Kites are generally considered less mobile and tied to more regularly flooded basins, but further research has demonstrated greater movement than expected. Not only has there been more localized shifting of populations but longer movements north into the St John's river system have been noted. Some workers feel that this has always been the natural pattern (Takekawa and Beissinger 1989), as it is in White Ibis and other waders (Bildstein *et al.* 1990).

9.6.7 WOODED WETLANDS

In areas where water-tolerant shrubs and trees become dominant species in the deeper basin wetlands of the northeastern USA, succession has been modeled on the basis of aerial photography spanning 20 to 33 years (Larson and Golet 1982), although many of these changes take much longer. Newer or more shallow wetlands have characteristics not unlike those elswhere in the world, but sedimentation and water manipulation result in a gradual shift from a dominance of meadow plants to shallow marshes, and then toward deep marsh plants and finally to shrub-dominated swamps. Ultimately, water-tolerant trees persist on wet, organic soils presumably formed within the wetland (Fig. 9.4). Based on bird studies in swamps (Aldrich 1945), such major changes in habitat structure would lead to changes from wetland birds to wet-forest species.

In the margins and backwaters of streams, wetland vegetation and change are dictated by the rates and timing of flooding events. A backwater shrub swamp in East Texas, once enclosed by dense bottomland hardwood forest, was opened by the catastrophic mortality of large overcup oak trees resulting from flooding events (Weller 1988). These tree ranged from 50 to 70 feet (15–21m) high and were over 100 years old. Although this oak species can tolerate flooding throughout much of the growing season, it is less stressed and growth is better when the water subsides with the advent of the growing season. The flooding may have been caused by water from several years of high rainfall and high creek levels and a small beaver dam. After the kill, the herbaceous vegetation flourished as it cannot do under the dark canopy and included typical wetland shallow-water or edge species, such as annual smartweeds, and seeded perennials such as grasses, sedges, and rushes. Concurrently, Water Elms, a deeper-water scrub species about 30 feet (9m) in height, found an ideal seed bed for its annual crop, and seedlings developed widely. These seedling rapidly became dominants that eventually will shade out the herbaceous species. The result is that the deeper-water vegetation zone has moved up slope – as well as the birds (Fig. 8.11).

9.6.8 STREAM-SIDE OR RIPARIAN WETLANDS

An example of stream modifications such as dam construction exists in northern Iowa where we followed vegetation periodically for nearly 30 years (Weller, Kaufmann and Vohs 1991). Damming upstream has had two major effects. First, pools formed behind several dams became rather sterile lakes with few marginal emergents and relatively few aquatic submergents. However, use by bay diving ducks and geese increased. Second, reduced downstream flow modified vegetation of temporary pools from diverse mudflat and marsh-edge annual plants to more perennial grasses characteristic of more shallow and less-flooded stream banks. Reversion is possible with time and return to traditional small-stream water regimes.

On a larger scale, the Platte River of Nebraska has been impacted by upstream dams in Colorado and Nebraska build for urban water supplies and agricultural irrigation, with the result that a traditional steady and shallow flow has been replaced by a reduced and less predictable flow. As a result, vegetation along the river and on sand and gravel islands has changed drastically toward drier streamside and woody types. Sparse herbaceous cover on islands has given way to willows and cottonwoods, completely changing the character of the wildlife habitat. As a result, shallow-water roosting islands are dry, vegetated, and unattractive to the Sandhill Cranes during early spring migration periods. Flooding of backwater and other marginal wetlands also has been reduced, thus drying favored feeding areas of cranes on tuber-producing sedges. During summer, the islands are exposed but are no longer habitat for Piping Plovers and Least Terns, which need open, sandy islands. Only restoration of the stream flow will allow potential recovery, and the area will then require management such as cutting of trees and brush, which is costly.

9.6.9 TIDAL ESTUARINE WETLANDS

Many people have concluded that coastal estuarine systems show less variability in plant communities on short time scales and are slower to change on long time scales because sea levels change slowly. But where tidal fluctuations are low, as in the Gulf of Mexico, estuarine wetlands experience dynamic water regimes, often as a product of the amount of freshwater inflow and local rainfall, and they can have a major impact on vegetation. The brackish water characteristic of estuary wetlands is a mix of freshwater and saltwater. The heavier saltwater sinks and is overridden by the lighter freshwater in a pattern termed the saltwater wedge. But water often enters the estuary as sheet flow and percolation everywhere along the coast, and blending of fresh- and saltwater is gradual. Where freshwater volume is high and continuous, the salinity is low. A reduction in rainfall or a major storm surge from the sea changes this ratio, and plants may be killed from the sudden high salinity, creating a germination site not unlike that of shallow, freshwater basin wetlands.

Seasonal variation in salinity and water depths (tide height) occurs as a result of shifts in lunar/sun gravitational pull, which creates different tide levels at different seasons (e.g., "spring tide"). Higher tidal areas well flooded early in the year often dry out in summer, and obviously their salt concentration increases – unless there is continuing freshwater inflow. That rainfall pattern varies by area of the country, and from year to year. Therefore, the more saline areas along the coast are not the areas regularly flooded by saltwater (which normally is less than 35 p.p.t.) but the irregularly flooded and seasonally drying areas that form "salt flats," where concentrations commonly exceed 50 p.p.t. This is an important influence on plant succession, and has important implications for restoration of wetlands lost along the coast for various reasons. These tidally or wind-influenced saltflats have their own avifauna, mainly because they also have a unique invertebrate "infauna," and such areas generally are unappreciated by non-ecologists.

At the San Bernard National Wildlife Refuge southwest of Houston, Texas, we have observed bird use of wetlands over three different summers reflecting wet, dry, and average years. The presence of a few inches of water in the areas above the tidal reach produces amazing differences in plant and bird communities, and the freshwater inflow has made one tidal creek near-fresh in some years and full-strength seawater in others. One might classify the area as estuarine in some years and freshwater (palustrine) wetland in others (Weller 1994).

Bird assemblages changed with the water and perhaps with the plants. During water declines, the more aquatic birds declined and presumably moved elsewhere, and the mudflats were invaded by shorebirds and more upland species. When water returned, so did the swimming and diving birds.

These examples all reflect the natural variability of wetlands over various time scales. Frankly, we are not very adept at dealing with such variation in processes over long time periods, or with the lack of constancy in the product. Most people expect wetlands to be constant. We expect them to have discrete edges and stay out of our fields and buildings. We expects plants and soils and animals to stay put and to have characteristics that will clearly delineate what is a wetland and what is terrestrial. We expect them to remain either fresh or salty. They do not. It is true that not all wetlands are as variable as the examples above. Among those are the more lake-like systems where the water is more available and the plant species tend to be perennial aquatics. Another more stable wetland type is the bog, which may float on the surface of a stable water body or may have water upwelling. In either case, saturation is nearly constant, and plants are probably more stable and predictable for bird use.

References

Aldrich, J. W. (1945). Birds of deciduous forest aquatic succession. *Wilson Bulletin* 57, 243–5.

Allen, A. A. (1914). The red-winged blackbird, a study in the ecology of a cat-tail marsh. *Proceedings of the Linnaean Society of New York* 24, 43–128.

Allison, S. K. (1992). The influence of rainfall variability on the species composition of a northern California salt marsh plant assemblage. *Vegetatio* 101, 145–60.

Beecher, W. J. (1942). *Nesting birds and the vegetative substrate.* Chicago, IL: Chicago Ornithological Society.

Begon, M., Harper, J. L., and Townsend, C. R. (1996). *Ecology*, 3rd edn. Oxford: Blackwell Science.

Bellrose, F. C., Paveglio, F. L. Jr, and Steffeck, D. W. (1979). Waterfowl populations and the changing environment of the Illinois River Valley. *Illinois Natural History Survey Bulletin* 32, 1–54.

Bertness, M. D. and Ellison, A. M. (1987). Determinants of pattern in a New England salt marsh plant community. *Ecological Monographs* 57, 129–47.

Bildstein, K. L., Post, W., Johnston, J., and Frederick, P. (1990). Freshwater wetlands, rainfall, and the breeding ecology of White Ibises in coastal South Carolina. *Wilson Bulletin* 102, 84–98.

Breeden, S. and Breeden, B. (1982). The drought of 1979–1980 at the Keoladeo Ghana Sanctuary, Bharatpur, Rajasthan. *Journal of the Bombay Natural History Society* 79, 1–37.

Briggs, S. V. (1992). Movement patterns and breeding adaptations of arid zone ducks. *Corella* 16, 15–22.

Browder, J. A. (1978). A modeling study of water, wetlands, and wood storks. In Research Report No. 7: *Wading birds,* ed. A. Sprunt, pp. 325–46. New York: National Audubon Society.

Caithamer, D. F. and Dubovsky, J. A. (1997). *Waterfowl population status, 1997.* Washington, DC: U S Fish & Wildlife Service.

Carpenter, S. R. (1981). Submersed vegetation: an internal factor in lake ecosystem succession. *American Naturalist* 118, 372–83.

Chabreck, R. H. (1988). *Coastal marshes: ecology and wildlife management.* Minneapolis, MN: University of Minnesota Press.

Crocker, W. (1938). Life span of seeds. *Botanical Review* 4, 235–72.

Crome, F. H. J. (1988). To drain or not to drain? – Intermittent swamp drainage and waterbird breeding. *Emu* 88, 243–48.

Custer, T. W., Hines, R. K., and Custer, C. M. (1996). Nest initiation and clutch size of Great Blue Heron on the Mississippi River in relation to the 1993 flood. *Condor* 98, 181–8.

Damman, A. W. H. and French, T. W. (1987). *The ecology of peat bogs of the glaciated Northeastern United States: a community profile.* Report 85(7). Washington, DC: US Fish and Wildlife Service.

Fredrickson, L. H. and Reid, F. A. (1990). Impacts of hydrologic alteration on management of freshwater wetlands. In *Management of dynamic ecosystems,* ed. J. M. Sweeney, pp. 71–90. West Lafayette, IN: The Wildlife Society North Central Section.

Frith, H. J. (1967). *Waterfowl in Australia.* Honolulu, HI: East-West Press

Golet, F. C. and Parkhurst, J. A. (1981). Freshwater wetland dynamics in South Kingstown, Rhode Island, 1939–1972. *Environmental Management* 5, 245–51.

Gosselink, J. G. (1984). *The ecology of delta marshes of coastal Louisiana: a community profile.* FWS/OBS-81/24. Washington, DC: US Fish & Wildlife Service.

Gunderson, L. H., Light, S. S., and Holling, C. S. (1995). Lessons from the Everglades. *BioScience* Supplement 1995: S66–S73.

Halse, S. A. and Jaensch, R. P. (1989). Breeding seasons of waterbirds in Southwestern Australia – the importance of rainfall. *Emu* 89, 232–49.

Harris, S. W. and Marshall, W. H. (1963). Ecology of water-level manipulations of a northern marsh. *Ecology* 44, 331–42.

Heitmeyer, M. E. and Fredrickson, L. H. (1981). Do wetland conditions in the Mississippi Delta hardwoods influence mallard recruitment? *Transactions of the North American Wildlife & Natural Resources Conference* 46, 44–57.

Kadlec, J. A. (1962). Effects of a drawdown on a waterfowl impoundment. *Ecology* 43, 267–81.

Kantrud, H., Krapu, G. L., and Swanson, G. A. (1989). *Prairie basin wetlands of the Dakotas: a community profile.* Biological Report 85(7.28). Washington, DC: U S Fish & Wildlife Service.

Kauppinen, J. and Vaisanen, A. (1993). Ordination and classification of waterfowl communities in south boreal lakes. *Finish Game Research* 48, 3–23.

Kruse, A. D. and Bowen, B. S. (1996). Effects of grazing and burning on densities and habitats of nesting ducks in North Dakota. *Journal of Wildlife Management* 60, 233–46.

Kushlan, J. A. (1989). Avian use of fluctuating wetlands. In *Freshwater wetlands and wildlife,* eds. R. R. Sharitz and J. W. Gibbons, pp. 593–604. Washington, DC: U S Department of Energy.

Larson, J. S. and Golet, F. C. (1982). Models of freshwater wetland change in southeastern New England. In *Wetlands ecology and management,* ed. B. Gopal, pp. 181–185. Jaipur, India: National Institute of Ecology and International Scientific Publications.

Meeks, R. L. (1969). The effect of drawdown date on wetland plant succession. *Journal of Wildlife Management* 33, 817–21.

Miller, M. W. and Nudds, T. D. (1996). Prairie landscape change and flooding in the Mississippi River Valley. *Conservation Biology* 10, 847–53.

Murkin, H. R. and Kadlec, J. (1986). Responses by benthic macroinvertebrates to prolonged flooding of marsh habitat. *Canadian Journal of Zoology* 64, 65–72.

Murkin, H. R., Kaminski, R. M. and Titman, R. D. (1982). Responses by dabbling ducks and aquatic invertebrates to an experimentally manipulated cattail marsh. *Canadian Journal of Zoology* 60, 2324–32.

Niering, W. A. and Warren, R. S. (1980). Vegetation patterns and processes in New England salt marshes. *BioScience* **30**, 301–7.

Penfound, W. T. and Hathaway, E. S. (1938). Plant communities in the marshlands of southeastern Louisiana. *Ecological Monographs* **8**, 1–56.

Pennings, S. C. and Callaway, R. M. (1992). Salt marsh plant zonation: the relative importance of competition and physical factors. *Ecology* **73**, 681–90.

Reimold, R. J., Linthurst, R. A., and Wolf, P. L. (1975). Effects of grazing on a salt-marsh. *Biological Conservation* **8**, 105–25.

Robel, R. J. (1961). Water depth and turbidity in relation to growth of Sago Pondweed. *Journal of Wildlife Management* **25**, 436–8.

Rogers, J. P. (1959). Low water and lesser scaup reproduction near Erickson, Manitoba. *Transactions of the North American Wildlife & Natural Resources Conference* **24**, 216–24.

Stanley, E. H., Fisher, S. G., and Grimm, N. B. (1997). Ecosystem expansion and contraction in streams. *BioScience* **47**, 427–35.

Takekawa, J. E. and Beissinger, S. R. (1989). Cyclic drought, dispersal, and the conservation of the snail kite in Florida: lessons in critical habitat. *Conservation Biology* **3**, 302–11.

van der Valk, A. G. (1981). Succession in wetlands: a Gleasonian approach. *Ecology* **62**, 688–96.

van der Valk, A. G. and Davis, C. B. (1978). The role of seed banks in the vegetation dynamics of prairie glacial marshes. *Ecology* **59**, 322–35.

van der Valk, A. G. and Welling, C. H. (1988). The development of zonation in fresh-water wetlands. In *Diversity and pattern in plant communities*, eds. M. J. During, M. J. A. Werger, and J. H. Willems, pp. 145–58. The Hague, Netherlands: SPH Academic Publishing.

Voigts, D. K. (1976). Aquatic invertebrate abundance in relation to changing marsh conditions. *American Midland Naturalist* **95**, 312–22.

Weller, M. W. (1988). Bird use of an east Texas shrub wetland. *Wetlands* **8**, 145–58.

Weller, M. W. (1994). Bird/habitat relations in a Texas estuarine marsh during summer. *Wetlands* **14**, 293–300.

Weller, M. W. and Fredrickson, L. H. (1974). Avian ecology of a managed glacial marsh. *Living Bird* **12**, 269–91.

Weller, M. W. and Spatcher, C. E. (1965). *Role of habitat in the distribution and abundance of marsh birds*. Special Report 43. Ames, IA: Iowa State University Agriculture and Home Economics Experiment Station.

Weller, M. W., Wingfield, B. H., and Low, J. B. (1958). Effects of habitat deterioration on bird populations of a small Utah marsh. *Condor* **60**, 220–6.

Weller, M. W., Kaufmann, G. W., and Vohs, P. A. Jr (1991). Evaluation of wetland development and waterbird response at Elk Creek Wildlife Management Area, Lake Mills, Iowa, 1961 to 1990. *Wetlands* **11**, 245–62.

Wiggins, G. B., MacKay, R. J., and Smith, I. M. (1980). Evolutionary and ecological strategies of animals in temporary pools. *Archiv für Hydrobilogie*, Supplement **58**, 97–206.

Wilson, S. D. and Keddy. P. A. (1985). Plant zonation on a shoreline gradient: physio-logical response curves of component species. *Journal of Ecology* **73**, 851–60.

10

Population consequences of wetland abundance and quality

The influence of wetland characteristics on bird populations and species composition is a subject of esthetic, economic, and scientific importance, and, therefore, the subject of continuing research efforts. If wetlands are variable in number, size, and quality through either short-term or long-term changes, we expect variation in their attraction to waterbirds, differential success in nesting, feeding, or roosting, and, ultimately, differences in the abundance and variety of bird species using them. Numerous biologists have reported data supporting this pattern for various species, but estimates are rarely clear-cut nor conclusions simple because of the many variables that influence population estimates on a local area – especially when dealing with long-range migrants and diverse species. As a result, we must be cautious in accepting what seem like obvious conclusions when causative influences are so difficult to separate.

Water is the chief driver and figures prominently in most studies, some of which have been mentioned above. A common and seemingly logical assumption is that stable wetlands should have the greatest diversity of bird species and the largest numbers because of the predictability of resources. However, most studies and several summaries of information on wetland productivity indicate that diversity and populations of bird species are greatest in those areas with unstable water regimes (Mitsch and Gosselink 1993, Weller 1995). Fluctuations in water or even drawdown periods allow oxidation and enhanced biochemical processes that induce productivity. Food production may subsequently be elevated, but even then, other factors like food availability or concentration influence bird use. It is not surprising that bird populations and bird diversity seem related to the level of eutrophication (Nilsson 1978). Stable wetlands probably have greater stability in population and species variance, but merely at considerably lower levels.

We must consider these influences at various temporal and spatial scales. Because many wetlands are not at the same water level or stage of succession or level of productivity year after year, we must assess production or survival of birds over a longer time span than for those species typically viewed as annually consistent. Short-lived birds like songbirds have few chances at reproduction during their lifetime and need to find suitable habitat for breeding if the

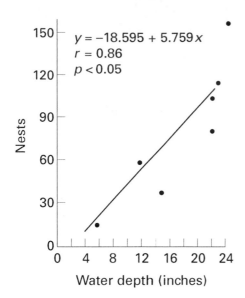

Figure 10.1. Nest numbers of Blue-winged Teal in relation to a water index for a wetland complex in northwest Iowa (Reproduced with permission from Weller (1979) *Journal of Wildlife Management* 43, 367–74. © The Wildlife Society.).

population is to survive. Local or long-range mobility is, therefore, essential to finding suitable habitat – summer or winter. Theoretically, long-lived birds could "sit-out" a year or more waiting for better conditions, but what are the risks and benefits of doing so? Some seem to occupy traditional areas but not reproduce, and others experience poor success because of predation and lack of nest attentiveness, as seen in breeding Lesser Scaup (Rogers 1959) and ducks, geese, and herons under drought conditions (Weller, Wingfield and Low 1958). Flood conditions can be just as devastating with loss of nests caused by wave action and poor cover, as shown for several grebes and American Coots (Boe 1993, Fredrickson 1970, Weller and Spatcher 1965). These observations suggest the magnitude of loss of species, populations, or nest success on a single wetland, but we recognize that wetlands differ and that all may not suffer the same level of impact. Therefore, if we consider these parameters of bird use at the level of a complex of wetlands, or a group of wetlands at some larger regional scale, we can expect quite different levels of response because there is less variability of resources in the entire group of wetlands. Most long-term studies have been of either single wetlands or a small complex and have shown dramatic variation in bird populations and species composition when wetlands are changing rapidly as a product of water regimes. Variation in populations from year to year commonly are correlated with water levels and are reflected in wetlands followed over several years (ducks, Fig. 10.1) (Weller 1979) and for different wetlands in the same area (shorebirds, Fig. 10.2) (Skagen and Knopf 1994).

Studies of year-to-year variation also can provide clues to features that constitute optimal conditions, or at least the extremes of suitability, as demonstrated by changes in species richness or species abundance under either

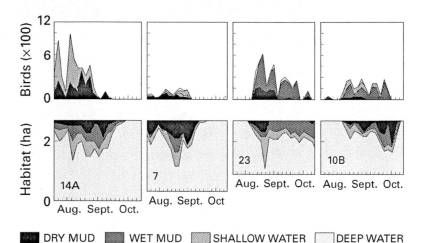

Figure 10.2. Influence of water regimes on shorebird numbers of four wetland microhabitats during migration periods on the Great Plains (Skagan and Knopf 1994, with permission).

extremely wet or dry conditions. Several bird species linked to prairie pothole edges also were included in grassland models by Cody (1985), who analyzed the relationships between rainfall variation and spring temperatures with bird density (Fig. 10.3)(Cody 1985). The positive effects of high rainfall is more conspicuous in the birds of the wetland edge, like Bobolinks, Red-winged Blackbirds, and ducks, than with more upland species that also nest in warmer periods. Dry periods affect especially the quality of basin wetlands for swimming and fish-eating birds like ducks and herons (Fig. 10.4) (Derksen and Eldridge 1980; Weller *et al.* 1958; Weller 1981). But the other extreme, flooding, may disrupt breeding even by swimming birds that nest over water like ducks, and sometimes without measureable increases in what seem like suitable adjacent areas (Foote 1989). The effect of floods on riverine wetland and floodplain birds has been studied less intensively, presumably because it is unpredictable. In the major Mississippi River flood of 1993, species richness and species abundance of floodplain birds declined after the flood either because of reduced habitat quality or because of poor reproductive success, but some water birds were less affected (Miller and Nudds 1996, Knutson and Klaas 1997). In this case, increases in abundance were noted in adjacent unflooded areas, presumably caused by displacement during the flood period. However, impacts on individual species vary with food resources and availability (Custer, Hines and Custer 1996).

Presence of breeding birds has been correlated with the number of water bodies in a landscape region for Canvasbacks (Fig. 10.5)(Sugden 1978) and for ducks by taxonomic groups (Johnson, Nichols and Schwartz 1992). Wetland formation by beavers is important in a number of areas of North America, and harvesting of beavers potentially could reduce the number of beaver ponds and thereby wetland bird habitat. However, studies of beaver harvest and duck populations suggest that minor beaver harvest is not significant but continued harvest over 3 to 4 years does result in reduced waterfowl populations (McCall

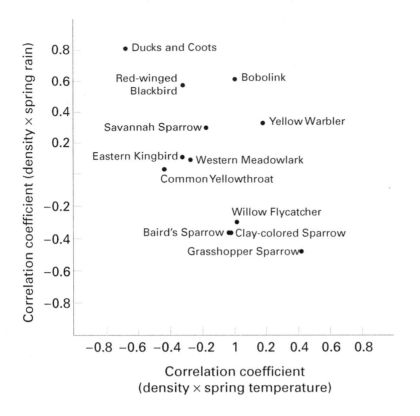

Figure 10.3. Relationship of various grassland and wetland birds to climatic regimes in the North American prairie (Cody 1985, with permission by Academic Press).

et al. 1996). Few studies seem to have been conducted on more stable wetlands where we might expect variation in numbers to be more a result of continent-wide population changes or changes in resources other than water.

Vegetation structure and food resources are inseparably linked to wetland selection or to within-wetland habitat selection on a breeding area. Returning adults are forced to abandon a former or potential breeding area if nesting cover and food are not available. Hence, risk of mobility and search would confront them even if nest or roost sites were available. In both American Coots and Yellow-headed Blackbirds, we have observed but been unable to quantify the arrival, temporary occupancy, and then sudden population decline that suggested that birds had tested quickly and moved elsewhere in search of better resources. Loss of food resources during nesting can cause abandonment of the nests and young. This is especially obvious in colonial species such as terns, egrets, and, probably, pelicans. Terns may remain and nest if at least some food is available but reproductive success as measured by number of young per pair of terns is related to food biomass in a sigmoid fashion (Suddaby and Ratcliffe 1997). Brown and Urban (1969) noted abandonment of 10000 nests of African Great White Pelicans caused by a collapse of the fish population. And Wood Storks often abandon nests when food resources are unavailable because of high water.

Population consequences

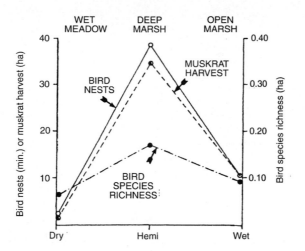

Figure 10.4. Bird species richness, nest numbers, and muskrat harvest under various water and vegetation in a single large wetland in northwest Iowa (Weller 1981:342, Figure 3).

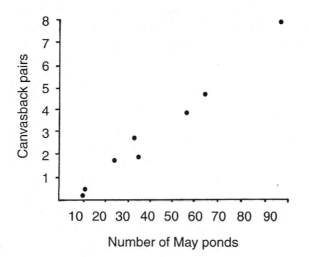

Figure 10.5. Numbers of Canvasback pairs established in eight wetland complexes in Saskatchewan during various water conditions (Sugden 1978).

In a single wetland, cover–water ratios and dispersion of vegetation influence richness and abundance of marsh birds (Fig. 10.6) (Weller and Fredrickson 1974). This pattern was tested experimentally for ducks (Kaminski and Prince 1981), which showed the highest densities in cover–water ratios of 50:50. Moreover, invertebrates also responded best to this ratio compared with the greater or smaller ratios of open water. Shoreline configuration is very important for breeding Trumpeter Swans (Banko 1960) and for ducks, presumably because of the isolation provided for territorial pairs of ducks, although Great-crested Grebes did not show the same response (Nilsson 1978). An area that requires study is the possibility of increased predation on or loss of eggs of birds nesting in small compared with large wetlands, as related to amount of edge and the source of predators as suspected in forest fragments

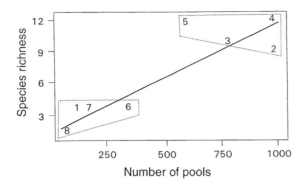

Figure 10.6. Regression of species richness against number of pools of open water in emergent marsh vegetation. The numbers represent observation points from different sample years, and the two clusters reflect low richness when few pools exist (i.e., dense vegetation) and high richness when vegetated is opened by interspersed pools (based on data in Weller and Fredrickson 1974).

(Wilcove 1985). This may be compensated in part by the protection of overwater nest sites, but many edge-nesters are exposed to greater snake and mammal predation.

Physical features like wetland size were discussed earlier and are important in explaining population size where species overlap to a limited degree by preferences for a particular wetland size. In the case of grebes, this is linked with vegetation: Horned Grebes use the smaller more open areas, Eared Grebes use medium to large areas that also lacked vegetation, and Pied-billed Grebes use medium to large areas that are more vegetated (Boe 1993, Faaborg 1976). Some extensive studies of Finish lakes with a wide range of nutrients, vegetation, and sizes has provided evidence of strong linkages of waterbird species to certain limnological characteristics such as (i) fish eaters in oligotrophic lakes (loons, mergansers); (ii) certain typical species in eutropic lakes (small grebes, coots, and teal, shovellers, and bay ducks); and (iii) species using shallow, nutrient-rich, but stressed lakes (termed mixotrophic), such as Eurasian Wigeon, Northern Pintail, and Tufted Duck (Kauppinen 1993).

Other physical features are important, often because they affect or are affected by water. Nesting White-throated Dippers in Britain seem to prefer lower stream gradients where currents are slower and pools are more common (Tyler and Ormerod 1994). This also seems true of feeding Spectacled Ducks of South America, which feed and nest along slower streams, whereas Torrent Ducks of the same areas of the Andes use smaller, higher streams with strong currents (Weller 1975). I suspect that feeding Blue Ducks of New Zealand fall somewhere between these extremes (Eldridge 1986), as do Harlequin Ducks of North America (Bellrose 1980).

Tidal regimes strongly affect shoreline exposure (Evans, Goss-Custard and Hale 1984), especially of mudflats, impacting the availability of feeding sites of many species, like shorebirds that feed on the wet mudflats of very shallow water (Fig. 10.7,) (Burger *et al.* 1977; Evans *et al.* 1984). Current telemetry methods have allowed precise measurement of spatial distribution and seasonal timing of mudflat use in Western Sandpipers and other small shorebirds

Population consequences

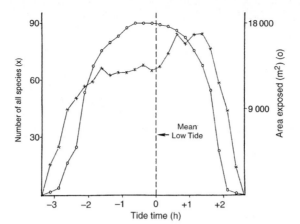

Figure 10.7. The influence of tidal regimes on exposed mud (O) and shorebird abundance (X) on an East Coast mudflat (Burger *et al.* 1977).

(Warnock and Takekawa 1995). Other carnivorous species such as Common Shelducks (Thompson 1982), Brown Teal (Weller 1974) and shorebirds (Rottenborn 1996) await low tide to feed or they move to other habitats elsewhere.

Foods are of obvious importance but are often difficult to relate to bird populations. Theoretically, birds that move widely and feed on well-distributed resources should select foraging areas with high density and availability of food. Moreover, nesting birds should have some mechanism for choosing potentially productive areas, other than merely returning to sites where they were successful earlier or where they were hatched. Some studies have shown a pattern of presence–absence relationships; for example, Snail Kites occur only where there are Apple Snails, whereas Hook-billed Kites are more flexible in foods, using Apple Snails in swamps and land snails in wet forest areas (Stiles and Skutch 1989). Carribean Flamingos seem to choose areas of dense food and food density declines with flamingo use, thereby influencing birds to move elsewhere (Arengo and Baldassarre 1995). Stream birds like dippers (Tyler and Ormerod 1994), Blue Ducks, and Torrent Ducks (Eldridge 1986) are most dense in areas of highest density of foods like mayflies and caddisflies. A strong positive relationship was shown in Australian wetlands between the density of midges (Chironomidae) and breeding efforts by ducks (Maher and Carpenter 1984). Goldeneye ducklings move overland at considerable risk to reach lakes highest in invertebrate food resources (Eriksson 1978). Territories of Common Shelducks are smaller and, therefore, their population is more dense in estuaries with adequate food levels; however, other factors seem to influence territory size and it is not strictly food-density dependent (Young 1970). Timing of laying and rearing have been closely linked to food availability in survival studies of dabbling, bay, and sea ducks (Bengtson 1971, Danell and Sjoberg 1977), and experimental marsh manipulation for invertebrates has demonstrated how dabbling ducks respond (Murkin, Kaminski and Titman 1982).

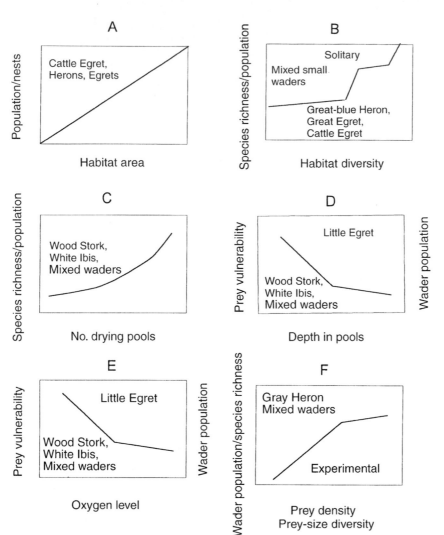

A

Population/nests | Habitat area

Cattle Egret, Herons, Egrets

B

Species richness/population | Habitat diversity

Solitary
Mixed small waders
Great-blue Heron, Great Egret, Cattle Egret

C

Species richness/population | No. drying pools

Wood Stork, White Ibis, Mixed waders

D

Prey vulnerability | Wader population | Depth in pools

Little Egret
Wood Stork, White Ibis, Mixed waders

E

Prey vulnerability | Wader population | Oxygen level

Little Egret
Wood Stork, White Ibis, Mixed waders

F

Wader population/species richness | Prey density Prey-size diversity

Gray Heron Mixed waders
Experimental

Figure 10.8. Relationships between habitat resources and use by waders generalized for application to restoration of the Kissimmee River Basin of Florida (Weller 1995). Use of two waterbird guilds as evaluation tools for the Kissimmee River restoration. *Restoration Ecology* 3(3), 211–14. Reprinted by permission of Blackwell Science, Inc.).

Impacts of sedimentation on water quality and abundance of submergent vegetation in oxbow lakes of the Illinois River Valley were reflected in populations crashes of American Coots and American Wigeon (Bellrose, Paveglio and Steffeck 1979). Several studies of shorebirds have shown correlations between their presence and food density and quality (especially size) of resources at the site (Bryant 1979, Myers, Williams and Pitelka 1980, Quammen 1984). Herons and egrets are affected by the density and the size of fish and other prey, which often are themselves affected by declining water levels and oxygen reduction, which increase vulnerability (Fig. 10.8) (Weller 1995). One experimental study of the Great White Heron of Florida (a white morph of the Great Blue Heron) showed that adding food increased clutch size and survival of young (Powell 1983). Increased heron and egret populations have been attributed to availability of foods from crayfish aquaculture in Louisiana (Fleury and Sherry 1995).

Food relationships also have been noted among more opportunistic or semiterrestrial species. Numbers of American Woodcock, which feed almost exclusively on earthworms in moist soils, show a strong relationship to earthworm density (Straw *et al.* 1995). Food abundance and food patch size both seem to influence the abundance and distance that Common Cranes will fly to feed (Alonso, Alonso and Veiga 1987). Recent observations on over-water aerial foraging on insects by Tree Swallows strongly suggests that they feed over food-rich wetlands, and that the additions of nutrients and the removal of fish may have enhanced the food availability of selected experimental ponds (McCarty 1997). These diverse examples should surprise no one, but they do reaffirm the absolute importance of food resources and the impacts that loss of such foods can have on populations of single species or on the diversity of bird species. From a conservation standpoint, reduced access to foods can only increase energy expenditures and induce exposure to other risks.

A factor tending to influence wetland suitability negatively is that ducks and fish may compete for the same foods. Mallard egg and clutch size seemingly were reduced in Swedish lakes which also contained fish (Pehrsson 1991). Foraging time (i.e., prey search time) was greater and weight and survival less for Mallard ducklings living in English gravel pits with higher fish populations, which also had lower invertebrate populations (Hill, Wright and Street 1987). Studies on Common Goldeneyes in Canada (Eadie and Keast 1982) are in agreement with studies in Sweden (Eriksson 1978) that fish and ducklings compete for similar foods.

Population response to new or improved habitat conditions is either rare or is rarely recognized for what it is. Resource-based increases in populations have been recorded regularly in field-feeding species. The availability of rice, sorghum, corn and similar grains resulting from expanding agriculture and from methods that leave extensive residual foods in fields seem responsible for the growth of populations and increased distribution of Lesser Snow Geese and White-fronted Geese (Bateman, Joanen and Stutzenbaker 1988), Black-bellied (Bolen and Rylander 1983) and Fulvous Whistling Ducks (Hohman, Stark and Moore 1996), and, probably, Sandhill Cranes (Krapu *et al.* 1984) in North America. Similar responses have been noted elsewhere with granivorous waterfowl such as Upland Geese in South America (Weller 1968) or grain-feeding ibises and godwits in Saudi Arabia (Rahmani and Shobrak 1992), but the lack of population data has not allowed direct linkage to the improved amount and availability of foods. As with other wetland types, avian adaptability in food use and mobility induce shifts to better feeding areas. When field-feeding conditions are poor, as with seasonal change (Hobaugh 1985), Lesser Snow Geese shift to traditional marsh foods such as tubers and rhizomes. Heavy damage to foods in staging areas caused by overgrazing of tubers seems to be affecting production of geese in Arctic Canada. Because survival of Lesser

Snow Geese has been so good in overwintering areas, increased numbers of adults returning to breeding grounds seem to be stressing the limited food supplies (Cargill and Jeffries 1984).

Some data from wintering areas suggest a relationship between breeding success and wintering conditions for the Mallard duck (Heitmeyer and Fredrickson 1981). Population variation during wet and dry years in the wintering areas infer that birds wintering under good food and water conditions are more successful at nesting in the following year. Similar relationships have been shown for breeding habitat, but density relationships complicate conclusions (Kaminski and Gluesing 1986). Nevertheless, these observations tend to support the logical conclusion that we cannot expect birds to breed at capacity when wintering, migration, or breeding habitats are poor. Nor can we expect populations and diversity to remain constant when the quality of the habitat is reduced naturally or by human influence.

At a geographical scale, climatic regimes are assumed to influence numbers and richness of breeding birds. We might logically assume that warm summers and freezing winters would influence wetland bird use. But there are so many influences on behavior, environmental factors, and geological history that any such effect would be gross. Superficially, species richness of breeding birds in wetlands in Texas, Iowa, North Dakota, and Alaska are not as variable as one might expect: there are about 22–28 species in most areas, but with great species differences resulting from adaptations for different microhabitats at various latitudes (e.g., use of mudflats or even uplands versus flooded emergents for feeding and nesting). Other population-related influences over time that are intertwined with habitats are the occurrence diseases like botulism and fowl cholera (Baldassarre and Bolen 1994). These are affected by water regimes, which may influence population density and disease spread. Fowl cholera currently is under intense study because it is unpredictable and potentially devastating, and perhaps linked to drought conditions on wintering areas. It has been especially prominent among Lesser Snow Geese of the Gulf Coast and the western USA, which involves long-distant migrants from Arctic Canada and Russia.

Diseases of food plants such as eelgrass have had major impacts on the numbers, distribution, condition, and behavior of Atlantic Black Brant in North America and Europe. The species is very dependent on eelgrass, an estuarine plant severely affected by a disease which in the 1930s wiped out major populations along the Atlantic Ocean (Cottam, Lynch and Nelson 1944). Moreover, its importance in the entire food web involving other birds, invertebrates, and fish demonstrates ecological community and system viewpoints during the early 1900s (Fig. 10.9). The causative organism of the eelgrass disease, a marine slime mold, has only recently been identified during a lesser recurrence in the 1980s. However, during its peak in the 1930s, as during cold spells that freeze over estuaries, Brant fed in urban parks and lawns on grass, as a temporary survival effort. Population declines were major and lasted many

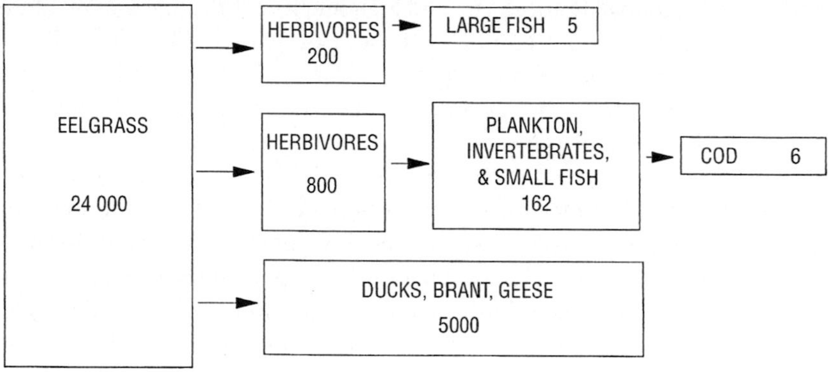

Figure 10.9. A semiquantitative model of eelgrass, fish, and waterfowl trophic relationships based on early studies of annual production of seagrasses in Denmark (numbers in 1000 metric tonnes) (redrawn from Thayer, Wolfe and Williams 1975, with permission).

years until recovery of the plant food. Recently, the western Pacific population has experienced major declines attributed to eelgrass losses. These are blamed on estuarine pollution and perhaps oyster mariculture, as well as boating disturbance that prevents the use of the food resources available (Einarsen 1965, Wilson and Atkinson 1995).

The above observations suggests a strong and logical relationship between the number, size, and quality of wetlands with species richness, population density, and community structure. But in our zeal to protect wetlands, it is common for conservationists to use the loss of wetlands (now over 50% in the contiguous USA according to Tiner (1984)) as a crude measure to estimate the decline in numbers – if not species – of birds in the USA since European settlement. But when we consider the fact that all wetlands vary in their suitability for birds, we would need an estimate based not only on the average number of wetlands available for use over the recent history of each species but also on their quality. The index that best approximates conditions for waterfowl breeding in the North American prairie potholes is an aerial survey where numbers (usually without measures of quality or size) of basin wetlands have been counted concurrently with duck populations. At this scale of tens of thousands of wetlands across the breeding range of the more common ducks, populations tend to decline after a series of years when more wetlands are dry, and enormous increases in populations can occur after a series of wet years when more basin wetlands contain water (Caithhamer and Dubovsky 1997, Cowardin and Blohm 1992, Dzubin 1969). There are many deviations from the expected, and a time lag of a year or more seems common owing to variation in breeding patterns and maturation of the various species.

This does not infer that the decline in ducks and other waterbirds over long periods is not proportional to wetland losses, but that we lack accurate background bird population data and information on wetland quality. Additionally, waterbirds are adaptable and movements mask some changes. Obviously, this also differs by bird group and life-history strategy: sparrows and wrens with small home ranges and high fidelity to breeding sites would be expected to suffer serious population losses even with declines of small wetlands. Ducks, rails, herons, and other waterbirds seemingly seek out better habitats (i.e., locally nomadic behavior induced by climatic conditions), and ducks and herons probably explore new breeding and wintering areas when pressed by poor conditions in previously used sites. But how far can the system be pushed? Most optimal habitat ranges have been populated, and marginal ones presumably can be used at some times but not at others because of variation in water regimes and climate. There are few opportunities to get this kind of information on wetland birds except perhaps on colonial breeders (e.g., ibises, egrets and herons), which are conspicuous by their large nesting colonies. This does not really make them easier to count because of layering of nests as well as tree cover, but they have been of concern because of dramatic declines and, therefore, are surveyed periodically. Several recent analyses of waders have demonstrated that colonies seem to shift geographically, i.e., abandoning long-used sites and appearing suddenly in other areas, states, or regions. We assume they move to better conditions, although we can only speculate on why. Water conditions are among the most important influences, with high water levels often as detrimental for some species as drought is for others because it reduces availability of prey (Kushlan 1976). Declines have been especially prominent in Florida and increases in Georgia and South Carolina; recently, increased numbers of species have been noted westward from Georgia to Louisiana and Texas concomitant with declines in Florida populations.

Another type of population change linked to wetland habitat quality or dynamics is the change in numbers of birds observed (or shot in the case of ducks and geese) at favoring migratory stopovers and wintering areas. One of the earliest of such observations resulted from kill records of the Winous Point Hunting Club along the shores of Lake Erie in Ohio, where Redhead ducks declined drastically from 1850 to 1900. Although this was cited as evidence of the serious decline of the species, I suspect it was a product of habitat change in the managed area, which favored dabbling ducks. Another was the case of the loss of the Canvasbacks at Lake Christina in Minnesota, which was attributed to increased turbidity and the decline of the favored submergent vegetation, sago pondweeds. Similar causes and impacts were documented in the Chesapeake Bay, where restoration efforts have included reduction of fertilizer and other nutrient flowing into the bay. Decline in water quality in pools of the Illinois River caused by upstream siltation and pollutants resulted in losses of invertebrate populations and pondweeds, with major impacts on use by migrant and wintering duck populations (Mills, Starrett and Bellrose 1966). Moreover, migratory routes have changed for Canvasbacks and perhaps for other plant-feeding ducks because they have shifted to other areas with more available resources.

Clearly, more data are needed on this important topic because it is the key

Plate IV. Breeding behavior and nesting sites.

(a) Male Yellow-headed Blackbird on a song perch of hardstem bulrush. The presence of robust emergent vegetation over water is essential to a successful territory, which attracts females that use prior-year's vegetation for building (Delta, Manitoba).

(b) Pair of Crested or Southern Screamers join in mutual display on a potential nest site of California Bulrush (General Lavalle, Argentina).

(c) Pair of Giant Canada Geese nesting on a muskrat lodge (Delta, Manitoba). Many large-bodied waterbirds that nest in high latitudes and have young requiring long growth periods initiate nesting during cold periods. Elsewhere in its original range, nesting on cliffs and in old raptor nests was common.

(*d*) Black Skimmers and other birds that feed in large open-water often use adjacent sandflats for nests, where exposure provides suitable winds for taking flight rapidly (Corpus Christi, Texas). They also will nest on wind-rowed dead vegetation in some areas, and artificial nest sites have been created.

(*e*) Floating nest of a Pied-billed Grebe made of wet emergent vegetation of the previous year. Eggs are normally covered by the incubating bird before departure (Ruthven, Iowa).

(*f*) Male Least Bittern covering young in a nest in cattail. Both sexes share incubation and feeding duties (Jewell, Iowa).

(*g*) Pair of Upland Sheldgeese that have led their brood from upland grassland to water for protection from potential predators (Falkland Islands).

(*h*) American White Pelican and young on nesting island in Chase Lake, North Dakota. Black coloration on the back of the head develops after eggs are laid and the birds crest is molted; at this stage, the bill knob typical of breeding birds also is lost.

(*i*) Nest hole of a Prothonotary Warbler in a Water Elm tree (Bethel, Texas).

(*j*) Colony of Cattle Egrets in Buttonbush shrub swamp (Bethel, Texas). Other species that nest in the same colony include Great and Snowy Egrets, Great Blue Herons, Anhingas, and White Ibis.

(*k*) Red Phalarope at a nest site in the sparse emergent vegetation of a tundra pond (Point Storkerson, Alaska).

Plate V. Resting and roosting.

(*a*) Resting and escape site on Bull Kelp used by Auckland Island Flightess Teal when disturbed by Antarctic Skuas and possibly by human intrusion (Adams Island).

(*b*) Multispecies daytime resting area for a mixed flock of terns and gulls, in a site where visibility and winds allow for rapid escape; vehicle disturbance causes evasive walking more commonly than flight (Caspian, Royal Forster's and Sandwich Terns, Laughing Gull, Port Aransas, Texas).

(*c*) Egyptian Sheldgoose resting on the back of a Hippopotamus, a site also used by egrets, oxpeckers, and, occasionally, sandpipers, which may feed on ticks and leeches there (Amboseli National Park, Kenya).

(*d*) An estuarine shallow-water rest site used by a mixed flock of waterbirds: Brown Pelican, Double-crested Cormorant, Wood Stork, Great and Snowy Egrets, Laughing Gull, and Grey Plover (Sanibel, Florida).

Plate VI. Wetland-related human activities.

(*a*) Informational sign to advise of the sensitivity of habitat and waterbirds to human intrusion (Corpus Christi, Texas).

(*b*) Boardwalk used to allow visitors to reach suitable bird viewing spots without serious impact on the wetland (Corpus Christi, Texas).

(*c*) People and waterbirds at St James' Park, London. This heavily used area is well known for its sizable population of Tufted Ducks and Mallards, which nest each year.

(*d*) Northern Shoveler Duck decoy from Italy made of cattail tied with wire. The crudely carved bill even mimics the spatulate feeding apparatus designed to take plankton.

(*e*) Goose decoy made of spruce twigs by Cree Indians of the James Bay Region of eastern Canada.

to the evaluation of conservation efforts for individual species and assemblages. Because of the great expense of conducting this kind of research, theoretical and mathematical models will be essential to synthesize the information we have and to use it to identify priorities in research. However, this can only work if we have better information on the strategies of resource use by birds, e.g., how and why bird use certain nest sites, what restricts use of roost sites, and how flexible birds are in changing migration routes.

References

Alonso, J. C., Alonso, J. A., and Veiga, J.P. (1987). Flocking in wintering Common Cranes: influence of population size, food abundance and habitat patchiness. *Ornis Scandinavia* 18, 53–60.

Arengo, F. and Baldassarre, G. A. (1995). Effects of food density on the behavior and distribution of nonbreeding American Flamingos in Yucatan, Mexico. *Condor* 97, 325–34.

Baldassare, G. A. and Bolen, E. G. (1994). *Waterfowl ecology and management.* New York: Wiley.

Banko, W. E. (1960). *The trumpeter Swan: its history, habits, and population in the United States.* North American Fauna 63. Washington, DC: U S Fish & Wildlife Service.

Bateman, H. A., Joanen, T., and Stutzenbaker, C. D. (1988). History and status of Midcontinent Snow Geese on their Gulf Coast winter range. In *Waterfowl in winter,* ed. M. W. Weller, pp. 495–515. Minneapolis, MN: University of Minnesota Press.

Bellrose, E C. (1980). *Ducks, geese and swans of North America* 3rd edn. Washington, DC: Wildlife Management Institute.

Bellrose, F. C., Paveglio, F. L. Jr and Steffeck, D. W. (1979). Waterfowl populations and the changing environment of the Illinois River Valley. *Illinois Natural History Survey Bulletin* 32, 1–54.

Bengtson, S. A. (1971). Variations in clutch-size in ducks in relation to the food supply. *Ibis* 113, 523–6.

Boe, J. S. (1993). Colony site selection by Eared Grebes in Minnesota. *Colonial Waterbirds* 16, 28–38.

Bolen, E. G. and Rylander, M. K. (1983). *Whistling ducks: zoogeography, ecology, anatomy.* Special Publication 20. Lubbock, TX: Texas Technical University Museum.

Brown, L. H. and Urban, E. K. (1969). The breeding biology of the Great White Pelican *Pelecanus onocrotalus roseus* at Lake Shale, Ethiopia. *Ibis* 111, 199–237.

Bryant, D. M. (1979). Effects of prey density and site character on estuary usage by overwintering waders (Charadrii). *Estuarine and Coastal Marine Science* 9, 369–84.

Burger, J., Howe, M. A., Hahn, D. C., and Chase, J. (1977). Effects of tide cycles on habitat selection and habitat partitioning by migrating shorebirds. *Auk* 94, 743–58.

Caithamer, D. F. and Dubovsky, J. A. (1997). *Waterfowl population status, 1997.* Washington DC: U S Fish & Wildlife Service.

Cargill, S. M. and Jeffries, R. L. (1984). The effects of grazing by lesser snow geese on the vegetation of a sub-Arctic salt marsh. *Journal of Applied Ecology* 21, 669–86.

Cody, M. L. (1985). Habitat selection in grassland and open country birds. In *Habitat selection in birds*, ed. M. L. Cody, pp. 191–226. Orlando, FL: Academic Press.

Cottam, C., Lynch, J. J., and Nelson, A. L. (1944). Food habits and management of American sea brant. *Journal of Wildlife Management* 8, 36–56.

Cowardin, L. M. and Blohm, R. J. (1992). Breeding population inventories and measures of recruitment. In *Ecology and management of breeding waterfowl*, eds. B. D. J. Batt, A. D. Afton, M. G. Anderson, C. D. Ankney, D. H. Johnson, J. A. Kadlec, and G. L. Krapu, pp.423–45. Minneapolis, MN: University of Minnesota Press.

Custer, T. W., Hines, R. K. and Custer, C. M. (1996). Nest initiation and clutch size of Great Blue Herons on the Mississippi River in relation to the 1993 flood. *Condor* 98, 181– 8.

Danell, K. and Sjoberg, K. (1977). Seasonal emergence of chironomids in relation to egg laying and hatching of ducks in a restored lake. *Wildfowl* 28, 129–35.

Derksen, D. V. and Eldridge, W. D. (1980). Drought-displacement of pintails to the Arctic Coastal Plain, Alaska. *Journal of Wildlife Management* 44, 224–9.

Dzubin, A. (1969). Comments on carrying capacity of small ponds for ducks and possible effects of density on Mallard production. In *Saskatoon wetlands seminar*, pp. 138–160. Report Series, No. 6. Ottawa: Canadian Wildlife Service.

Eadie, J. M. and Keast, A.. (1982). Do Goldeneye and perch compete for food? *Oecologia* 55, 225–30.

Einarsen, A. S. (1965). *Black brant, sea goose of the Pacific coast.* Seattle: University of Washington Press.

Eldridge, J. L. (1986). Territoriality in a river specialist, the Blue Duck. *Wildfowl* 37, 123–35.

Eriksson, M. O. G. (1978). Lake selection by Goldeneye ducklings in relation to the abundance of food. *Wildfowl* 37, 81–5.

Evans, P. R., Goss-Custard, J. D., and Hale, W. G. (eds.) (1984). *Coastal waders and wildfowl in winter.* Cambridge: Cambridge University Press.

Faaborg, J. (1976). Habitat selection and territorial behavior of the small grebes of North Dakota. *Wilson Bulletin* 88, 390–9.

Fleury, E. and Sherry, T. W. (1995). Long-term population trends of colonial wading birds in the southern United States: the impact of crayfish aquaculture on Louisiana populations. *Auk* 112, 758–61.

Foote, A. L. (1989). Response of nesting waterfowl to flooding in Great Salt Lake wetlands. *Great Basin Naturalist* 49, 614–17.

Fredrickson, L. H. (1970). The breeding biology of American coots in Iowa. *Wilson Bulletin* 82, 445–57.

Heitmeyer, M. E. and Fredrickson, L. H. (1981). Do wetland conditions in the Mississippi Delta hardwoods influence mallard recruitment? *Transaction of the North American Wildlife and Natural Resource Conference* 46, 44–57.

Hill, D., Wright, R., and Street, M. (1987). Survival of Mallard ducklings *Anas platyrhynchos* and competition with fish for invertebrates in a flooded gravel quarry in England. *Ibis* 129, 159–67.

Population consequences

Hobaugh, W. C. (1985). Body condition and nutrition of Snow Geese wintering in southeastern Texas. *Journal of Wildlife Management* **49**, 1028–37.

Hohman, W. L., Stark, T. M., and Moore, J. L. (1996). Food availability and feeding preferences of breeding Fulvous-whistling Ducks in Louisiana ricefields. *Wilson Bulletin* **108**, 137–50.

Johnson, D. H., Nichols, J. D. and Schwartz, M. D. (1992). Population dynamics of breeding waterfowl. In *Ecology and management of breeding waterfowl*, eds. B. D. J. Batt, A. D. Afton, M. G. Anderson, C. D. Ankney, D. H. Johnson, J. A. Kadlec, and G. L. Krapu, pp. 446–85. Minneapolis, MN: University of Minnesota Press.

Kaminski, R. M. and Gluesing, E. A. (1986). Density- and habitat- related recruitment in mallards. *Journal of Wildlife Management* **51**, 141–8.

Kaminski, R. M. and Prince, H. H. (1981). Dabbling duck and aquatic vertebrate response to manipulated habitat. *Journal of Wildlife Management* **45**, 1–15.

Kauppinen, J. (1993). Densities and habitat distribution of breeding waterfowl in boreal lakes in Finland. *Finish Game Research* **48**, 24–45.

Knutson, M. G. and Klaas, E. E. (1997). Declines in abundance and species richness of birds following a major flood on the Upper Mississippi River. *Auk* **114**, 367–80.

Krapu, G. L., Facey, D. E. Fritzell, E. K., and Johnson D. H. (1984). Habitat use by migrant Sandhill Cranes in Nebraska. *Journal of Wildlife Management* **48**, 407–17.

Kushlan, J. A. (1976). Feeding ecology and prey selection in the White Ibis. Condor **81**, 376–89.

Maher, M. and Carpenter, S. M. (1984). Benthic studies of waterfowl breeding habitat in South-western New South Wales. II. Chironomid populations. *Australian Journal of Marine & Freshwater Research* **35**, 97–110.

McCall, T. C., Hodgman, T. P., Diefenbach, D. R., and Owen, R. B. Jr (1996). Beaver populations and their relation to wetland habitat and breeding waterfowl in Maine. *Wetlands* **16**, 163–72.

McCarty, J. P. (1997). Aquatic community characteristics influence the foraging patterns of tree swallows. *Condor* **99**, 210–13.

Miller, M. W. and Nudds, T. D. (1996). Prairie landscape change and flooding in the Mississippi River Valley. *Conservation Biology* **10**, 847–53.

Mills, H. B., Starrett, W. C., and Bellrose, F. C. (1966). Man's effect on the fish and wildlife of the Illinois River. *Illinois Natural History Survey Biological Notes* **57**, 1–24

Mitsch, W. J. and Gosselink, J. G. (1993). *Wetlands*, 2nd edn. New York: Van Nostrand Reinholdt.

Muehistein, L. K. (1989). Perspectives on the wasting disease of eelgrass *Zostera marina*. *Diseases of Aquatic Organisms* **7**, 211–21.

Murkin, H. R., Kaminski, R. M., and Titman, R. D. (1982). Responses by dabbling ducks and aquatic invertebrates to an experimentally manipulated cattail marsh. *Canadian Journal of Zoology* **60**, 2324– 32.

Myers, J. P., Williams, S. L., and Pitelka, F. A. (1980). An experimental analysis of prey availability for Sanderlings. (Aves: Scolopacidae) feeding on sandy beach crustaceans. *Journal of Zoology* **58**, 1564–74.

Nilsson, L. (1978). Breeding waterfowl in eutrophicated lakes in south Sweden. *Wildfowl* **29**, 101–10.

Pehrsson, O. (1991). Egg and clutch size in the mallard as related to food quality. *Canadian Journal of Zoology* **69**, 156–62.

Powell, G. V. N. (1983). Food availability and reproduction of Great Blue White Herons, *Ardea herodias*: a food additive study. *Colonial Waterbirds* **6**, 139–47.

Quammen, M. L. (1984). Predation by shorebirds fish and crabs on invertebrates on intertidial mudflats: an experimental test. *Ecology* **65**, 529–37.

Rahmani, A. R. and Shobrak, M. Y. (1992). Glossy ibises (*Plegadis fascinellus*) and Black-tailed Godwits (*Limosa limosa*) feeding on sorghum in flooded fields in southwestern Saudi Arabia. *Colonial Waterbirds* **15**, 239–40.

Rogers, J. (1959). Low water and lesser scaup reproduction near Erickson, Manitoba. *Transactions of the North American Wildlife and Natural Resource Conference* **44**, 114–26.

Rottenborn, S. C. (1996). The use of coastal agricultural fields in Virginia as foraging habitat by shorebirds. *Wilson Bulletin* **108**, 783–96.

Skagen, S. K. and Knopf, F. L. (1994). Migrating shorebirds and habitat dynamics at a prairie wetland complex. *Wilson Bulletin* **106**, 91–105.

Stiles, F. G. and Skutch, A. F. (1989). *A guide to the birds of Costa Rica*. Ithaca, NY: Cornell University Press.

Straw, J. A. Jr, Krementz, D. G., Olinde, M. W., and Sepek, G. F. (1995). American Woodcock. In *Game bird management in North America*, eds. T. C. Tacha, and C. E. Braun, pp. 97–114. Washington, DC: International Association of Fish & Wildlife Agencies.

Suddaby, D. and Ratcliffe, N. (1997). The effects of fluctuating food availability on breeding Arctic Tern (*Sterna paradisea*). *Auk* **114**, 524–30.

Sugden, L. G. (1978). Canvasback habitat use and production in Saskatchewan parklands. *Canadian Wildlife Service Occasional Papers* **34**, 1–30.

Thayer, G. W., Wolfe, D. A., and Williams, R. B. (1975). The impact of man on sea grass systems. *American Scientist* **63**, 288–96.

Thompson, D. B. A. (1982). The abundance and distribution of intertidal invertebrates, and an estimate of their selection by Shelduck. *Wildfowl* **33**, 151–8.

Tiner, R. W. Jr (1984). *Wetlands of the United States: current status and recent trends*. Newton Corner, MA: US Fish and Wildlife Service.

Tyler, S. and Ormerod, S. (1994). *The dippers*. San Diego, CA: Academic Press.

Warnock, S. E. and Takekawa, J. Y. (1995). Habitat preferences of wintering shorebirds in a temporally changing environment: Western Sandpipers in the San Francisco Bay. *Auk* **112**, 920–30.

Weller, M. W. (1964). Distribution and migration of the redhead. *Journal of Wildlife Management* **28**, 64–103.

Weller, M. W. (1968). Notes on some Argentine anatids. *Wilson Bulletin* **80**, 189–212.

Weller, M. W. (1974). Habitat selection and feeding patterns of Brown Teal (*Anas castanea chlorotis*) on Great Barrier Island. *Notornis* **21**, 25–35.

Weller, M. W. (1975). Habitat selection by waterfowl of Argentine Isla Grande. *Wilson Bulletin* **87**, 83–90.

Weller, M. W. (1979). Density and habitat relationships of blue- winged teal nesting in northwest Iowa. *Journal of Wildlife Management* **43**, 367–74.

Weller, M. W. (1981). Estimating wildlife and wetland losses due to drainage and other perturbations. In *Selected proceedings of the Mid-west conference on wetland values and management*, ed. B. Richardson, pp. 337–46. St. Paul, MN: Minnesota Water Planning Board.

Weller, M. W. (1995). Use of two waterbird guilds as evaluation tools for the Kissimmee River restoration. *Restoration Ecology* 3, 211–24.

Weller, M. W. and Fredrickson, L. H. (1974). Avian ecology of a managed glacial marsh. *Living Bird* 12, 269–91.

Weller, M. W. and Spatcher, C. E. (1965). *Role of habitat in the distribution and abundance of marsh birds*, pp. 1–31. Special Report No. 43. Ames IA: Iowa State University Agriculture & Home Economics Experiment Station.

Weller, M. W., Wingfield, B. H., and Low, J. B. (1958). Effects of habitat deterioration on bird populations of a small Utah marsh. *Condor* 60, 220–6.

Wilcove, D. S. (1985). Nest predation in forest tracts and the decline of migratory songbirds. *Ecology* 66, 1212–14.

Wilson, U. W. and Atkinson, J. B. (1995). Winter and spring-staging use at two Washington coastal areas in relation to eelgrass abundance. *Condor* 97, 91–8.

Young, C. M. (1970). Territoriality in the Common Shelduck *Tadorna tadorna*. *Ibis* 112, 330–5.

Zwarts, L. and Warnink J. (1984). How Oystercatchers and Curlews successively deplete clams. In *Coastal waders and wildfowl in winter,* eds. P. R. Evans, J. D. Goss-Custard, and W. G. Hale, pp. 69–83. Cambridge: Cambridge University Press.

Further reading

Barnhart, R. A., Boyd, M. J., and Pequegnat, J. E. (1992). *The ecology of Humboldt Bay, California: an estuarine profile.* Biology Report 1. Washington, DC: U S Fish & Wildlife Service.

Blancher, P. J. and McNicol, D. K. (1991). Tree swallow diet in relation to wetland acidity. *Canadian Journal of Zoology* 69, 2629–37.

Draulans, D. (1987). The effect of prey density on foraging behaviour and success of adult and first-year Grey Herons (Ardea *cinerea*). *Journal of Animal Ecology* 56, 479–93.

Goss-Custard, J. D. (1979). Effect of habitat loss on the numbers of overwintering shorebirds. *Studies in Avian Biology,* 2, 167–77.

Henry, W. G. (1980). *Populations and behaviour of black brant at Humboldt Bay, California.* MSc Thesis. Arcata, CA: Humboldt State University.

Kersten, M., Britton, R. H. Dugan, P., and Hafner, H. (1991). Flock feeding and food intake in little egrets: the effects of prey distribution and behaviour. *Journal of Animal Ecology* 60, 241–52.

Kushlan, J. A. (1976). Feeding ecology and prey selection in the White Ibis. *Condor* 81, 376–89.

Moser, M.E. (1986). Prey profitability for adult Grey Herons *Ardea cinerea* and constraints on prey size when feeding nestlings. *Ibis* 128, 392–405.

Owen, M. and Black, J. M. (1990). *Waterfowl ecology.* Glasgow: Blackie.

Quammen, M. L. (1981). Use of enclosures in studies of predation by shorebirds on intertidal mudflats. *Auk* 82, 812–17.

Renfrow, D. H. (1993). The effect of fish density on wading birds' use of sediment ponds on an East Texas coal mine. Master's thesis. College Station, TX: Texas A&M University.

Schaffer, W. M. (1974). Optimal reproductive effort in fluctuating environments. *American Naturalist* **108**, 783–90.

Utschick, H. (1976). Die Wasservogel als Indikatoren für den okologischen Zustand von Seen (with English summary). *Verhandlungen der Ornithologischen Gesellschaft in Bayern* **22**, 395–438.

Weller, M. W. (1994). Seasonal dynamics of bird assemblages in a Texas estuarine wetland. *Journal of Field Ornithology* **65**, 388–401.

11

How birds influence wetlands

Although our major goal throughout the book has been to relate bird habitat use to wetland resources and features, birds that can directly affect a wetland ecosystem by their influence on vegetation and thereby bird habitat have an impact on other birds directly or alter ecosystem processes that may induce change within the entire system (Table 11.1).

Because of their mobility, birds facilitate movement of invertebrates to established wetlands that have lost their population via drought (Swanson 1984), as all invertebrates do not have eggs that can survive long periods without water (Wiggins, MacKay and Smith 1980). Birds also transport invertebrate eggs to newly created wetlands such as strip-mine ponds or river-formed wetlands. Common herbivorous crustaceans like amphipods have been especially prominent in studies of the plumage of hunter-killed or live waterfowl in both Europe and North America, but snails, snail eggs, and other macrocrustaceans have also been noted in plumage. One cannot wade in wetlands without experiencing invertebrates and seeds clinging to boots, so it is no surprise that birds (and also beavers and muskrats) carry them as well. Plant materials such as seeds and foliage of submergent plants are commonly found in or on the plumage of herbivorous ducks and coots. One study reported seeds of 12 saltmarsh plants, most of which had special adhesive devices or systems, carried by ducks (Vivian-Smith and Stiles 1994). This helps to explain a gradual increase in richness that occurs over time in wetland developed on strip-mined lands, where plant "seed banks" are minimal (McKnight, 1992). In such cases, wetland seed banks from 5- to 7-year-old ponds are commonly used to establish vegetation in newly created wetlands. Although some seeds may be wind-carried (willows, cattail, or fine seeds like sedges), most seeds are unlikely to be moved in this way. Birds also transport algae (Proctor 1962), seeds (DeVlaming and Proctor 1968), and invertebrates (Malone 1965) in their digestive tracts. Planktonic algae are passed through intact, although viability is doubtful for most (Atkinson 1971), but about 3% viability has been noted in seeds found in duck digestive tracts (Powers, Noble and Chabreck 1978). Although this figure seems small, it would only take a few seeds in the suitable sites used by ducks to establish a plant species.

The second major way that birds affect wetlands is as users. Herbivores like

Table 11.1 *How birds influence wetland development and community structure*

Seed dispersal: locally and globally

Invertebrate dispersal: locally and globally

Herbivore impacts on plant foliage, tubers, and rhizomes

Basin deepening by herbivores that dig out tubers

Nutrient deposition to water column and substrate that influence nutrient base and ecosystem processes

Consumption of invertebrates

Consumption of fish

Consumption of herbivores by predatory birds (hawks, waders)

Consequences of predation on other controlling species in the system

Lesser Snow Geese in fresh or saline marshes (Glazener 1946, Smith and Odum 1983), and Greylag Geese (Kvet and Hudek 1971) and White-fronted Geese in fresh marshes (personal observations) can be as devastating to even robust emergent vegetation as are Muskrats, Nutria, and feral pigs. In this way, they influence the many aspects of plant succession mentioned earlier and thereby have ecosystem-level consequences. Brant, swans, wigeon, and some bay ducks feed heavily on eelgrass foliage, although the impact is often modest and difficult to measure (Nienhuis and van Ierland 1978). Many duck species and coots in both coastal and inland saline settings feed heavily on the more delicate foliage of widgeongrass, presumably to the point of stripping areas and inducing birds to move elsewhere. Coots measurably impact *Hydrilla* sp., a favored, exotic submergent that grows in deep water and reaches to the surface (Esler 1989). Redhead ducks concentrate on and impact the abundance of rhizomes and tubers of seagrasses in winter (Cornelius 1977), and Sandhill Cranes eat sedge tubers in meadows during migration and breeding (Reineke and Krapu 1986). In all cases, their flock sizes and density at feeding sites suggest that they must have measurable consequences on the density and distribution of plant propagules. African flamingos, known for their concentration, which numbers in the hundreds of thousands, reportedly devour tons of planktonic organisms such as algae or invertebrates per day in interior saline lakes. While these levels of consumption are impressive, their impact on the system probably is more seasonal than permanent, but the long-term consequences of movement of this nutrient base away from the lake via nomadic or migratory movements requires detailed study.

Carnivorous birds remove major biomass in the form of invertebrates, fish, small mammals, and birds, and several meaningful estimates have been made. Studies of the use of mudflat invertebrate in fauna have been made in connection with the shorebird studies mentioned above (e.g., Quammen 1984). By use of captive Red Phalaropes, both reduction and size selection of crustacean prey has been suggested in Arctic wetlands (personal communication from S. I. Dodson). Herons and other fish-eating birds undoubtedly take enormous numbers of fish, but much of this feeding seems to be most efficient when water levels decline, oxygen is reduced, or fish are stunned by cold. An estimated take of 76% of the fish of a declining pond was noted compared with a 93% loss during total drying (Kushlan 1976). However, Smith *et al.* (1986) were unable to demonstrate significant diving duck impact on most invertebrates except for clams. Studies of oystercatchers, mentioned above also suggest that impact on large and relatively sedentary prey items may be greatest, but major reduction of the invertebrate community has not yet been demonstrated.

Mixed flocks of cormorants and pelicans do take many large fish by their efficient cooperative feeding efforts, but the real impact on the fish populations is very difficult to assess. Mergansers (and other sea ducks) that use eggs (i.e. spawn) in streams or in freshwater inflows to bays have been controlled in various places because of their potential competition with what are viewed as human resources. Consumption of small or young birds, herptiles, and mammals by predatory birds such as hawks or herons obviously are an established part of the food chain, moving energy from low to higher consumer levels.

Waterbird predation may influence ecosystem-level processes via impact on major herbivores that directly influence the vegetation base for all consumers. Several examples of heavy predation have been mentioned, but such impacts are difficult to assess in complex systems with many layers of consumers. In a more simple system, the role of sea ducks (eiders) on sea urchins, which feed on kelp has been investigated, and although measurable consequences were complicated by structural aspects of the vegetation and the technical difficulty of assessing foods in kelp "forests," strong evidence was obtained that eiders can affect sea urchin populations in small patches, and thereby potentially influence the survival of patches of kelp (Bustnes and Lonne 1995).

Less direct but perhaps more widespread influences of waterbirds on wetland occurs through nutrients added to the water column from defecation and sometimes from carcasses of flocking birds in breeding colonies and roosting sites. Among egrets and herons, this could have longer-term implications for the plant community, as mortality of trees has been attributed to heavy fecal deposition. Interestingly, dead trees are probably better nest sites for most herons and egrets, and the opening in the canopy must create new plant germination sites that ultimately influence local plant succession. Franklin's Gulls concentrate in enormous breeding colonies and potentially can have a major nutrient input to marsh emergents (McColl and Burger 1976) as can blackbirds and starlings roosting in cattail marshes (Hayes and Caslick 1984). Flocks of free-flying but site-attracted waterbirds have been shown to deposit significant nutrients with the potential to influence algal growth, thus entering the nutrient cycle, for example Canada Geese in Michigan (Manny, Wetzel and Johnson 1975), mixed species of ducks and coots in New York (Have 1973), marine birds in estuaries (Ganning and Wulff 1969), and Redhead

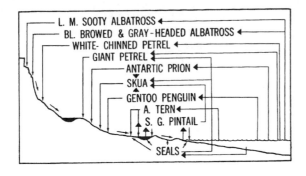

Figure 11.1. Hypothetical influence of hillside nesting of Antarctic seabirds on nutrients of lower freshwater wetlands that are rich in fairy shrimp and other invertebrates and are used for breeding by South Georgia (S. G.) Pintails (L. M., Light-mantled; BL., Black; A., Antarctic) (Weller 1980, with permission).

Ducks in coastal ponds in Texas (personal communication from M. K. Skorupa). Such waterbird concentrations may occur naturally from nesting, watering, and roosting sites but also result from human actions such as the attraction and feeding of wild or captive ducks on urban park lakes. Natural pollutants of the type described most noticeably contribute to eutrophy of shallow wetlands; often this is obvious in algal concentrations. In the more severe climate of an Antarctic island, wetland enrichment from seabirds nesting in and on cliffs seemingly results from the transport of nutrients derived from fish and large invertebrates, which are brought to land as food for nestlings (Fig.11.1).

References

Atkinson, K. M. (1971). Further experiments in dispersal of phytoplankton by birds. *Wildfowl* 22, 98–9.

Bustnes, J. O. and Lonne, O. J. (1995). Sea ducks as predators on sea urchins in a northern kelp forest. In *Ecology of fjords and coastal waters,* eds. H. R. Skjoldal, C. Hopkins, K. E. Erikstad, and H. P. Leinaas, pp. 599–608. New York: Elsevier Science.

Cornelius, S. E. (1977). Food and resource utilization by wintering redheads on lower Laguna Madre. *Journal of Wildlife Management* 41, 374–85.

DeVlaming, V. and Proctor, V. W. (1968). Dispersal of aquatic organisms: viability of seeds recovered from the droppings of captive killdeer and mallard ducks. *American Journal of Botany* 55, 20–6.

Esler, D. (1989). An assessment of American Coot herbivory of hydrilla. *Journal of Wildlife Management* 53, 1147–9.

Ganning, B. and Wulff, F. (1969). The effects of bird droppings on chemical and biological dynamics in brackish water rock pools. *Oikos* 20, 274–89.

Glazener, W. C. (1946). Food habits of wild geese on the Gulf Coast of Texas. *Journal of Wildlife Management* 10, 322–29.

Have, M. R. (1973). Effects of migratory waterfowl on water quality at the Montezuma National Wildlife Refuge, Seneca County, New York. *US Geological Survey Journal of Research* 1, 725–34.

Hayes, J. P. and Caslick, J. W. (1984). Nutrient deposition in cattail stands by communally roosting blackbirds and starlings. *American Naturalist* 112, 332–31.

Kushlan, J. A. (1976). Wading bird predation in a seasonal fluctuating pond. *Auk* 93, 464–76.

Kvet, J. and Hudec, K. (1971). Effects of grazing by Grey-lag Geese on reedswamp communities. *Hidrobiologia* 12, 351–9.

Malone, C. R. (1965). Dispersal of aquatic gastropods via the intestinal tract of water birds. *Nautilus* 78, 135–9.

Manny, B. A., Wetzel, R. G., and Johnson, W. C. (1975). Annual contribution of carbon, nitrogen, and phosphorus by migrant Canada Geese to a hardwater lake. *Verhandlungen Internationaler Vereinigung Limnologie* 19, 945–51.

McColl, J. G. and Burger, J. (1976). Chemical inputs by a colony of Franklin's Gulls nesting in cattail. *American Naturalist* 96, 270–80.

McKnight, S. K. (1992). Transplanted seed bank response to drawdown time in a created wetland in East Texas. *Wetlands* 12, 79–90.

Nienhuis, P. H. and van Ierland, E. T. (1978). Consumption of eelgrass, *Zostera marina*, by birds and invertebrates during the growing season in Lake Grevelingen (SW Netherlands). *Netherlands Journal of Sea Research* 12, 180–94.

Powers, K. D., Noble, R. E., and Chabreck, R. H. (1978). Seed distribution by water-fowl in southwestern Louisiana. *Journal of Wildlife Management* 42, 598–605.

Proctor, V. W. (1962). Availability of *Chara* oospores taken from migratory water-birds. *Ecology* 43, 528–9.

Quammen, M. L. (1984). Predation by shorebirds, fish and crabs on invertebrates on intertidal mudflats: an experimental test. *Ecology* 65, 529–37.

Reineke, K. J. and Krapu, G. L. (1986). Feeding ecology of sandhill cranes during spring migration in Nebraska. *Journal of Wildlife Management* 50, 71–9.

Smith, L. M., Vangilder, L. D., Hoppe. R. T., Morreale, S. J., and Brisbin, I. L. Jr (1986). Effects of diving ducks on benthic food resources during winter in South Carolina, USA *Wildfowl* 37, 136–41.

Smith III, T. J. and Odum, W. E. (1983). The effects of grazing by snow geese on coastal salt marshes. *Ecology* 62, 98–106.

Swanson, G. A. (1984). Dissemination of amphipods by waterfowl. *Journal of Wildlife Management* 48, 988–91.

Vivian-Smith, G. and Stiles, E. W. (1994). Dispersal of salt marsh seeds on the feet and feathers of waterfowl. *Wetlands* 14, 316–19.

Weller, M. W. (1980). *The island waterfowl.* Ames, IA: Iowa State University Press.

Wiggins, G. B., MacKay, R. J., and Smith, I. M. (1980). Evolutionary and ecological strategies of animals in temporary pools. *Archives für Hydrobilogie*, Supplement 58, 97–206.

12

Conservation implications

Wetland birds differ widely in their species composition and relative abundance within a community, but we lack thorough studies of communities in many wetland types and at various temporal scales. In addition to helping us to understand how such systems function, these data provide a basis for measuring change and evaluating impacts of natural and societal influences and, therefore, are important in conservation efforts. The purpose of this chapter is to summarize some of the patterns of habitat use by wetland birds and their potential biological properties, outlined above, and generalize on their relationship to conservation and management. Conserving wetland birds also requires an understanding of their relationships to society, because human populations do not have an impact on all groups of wetland birds uniformly (Weller 1988a). Moreover, some wetland types are more at risk than others because of alternative human uses, which will have a secondary effect on the success of birds there (Krapu *et al.* 1997).

Interest in wetland bird conservation often has focused on single species or taxonomic groups that have been of interest or importance for esthetic, recreational, or food values, and efforts to maintain populations have been funded by user-fees (e.g., hunting, bird watching). The use of general tax appropriations is less common except for those species that are especially rare; this has been an area of increasing concern because of the role that society has played in extirpations and extinctions. Although initial efforts at conservation of such species were aimed at legal protection, conservation strategies now tend to focus on preservation of suitable habitat or the assurance of continued functioning of existing habitat as the first step in protection of threatened populations, species, and communities.

The following concepts and commentaries are divided by general topics and include some issues and strategies of importance in wetland bird conservation. These are examples rather than a complete "cook book" of approaches. As with all generalizations, there may be major variation among bird groups, types of wetland, and geographical location, but they should help to identify patterns applicable in the development and evaluation of a conservation strategy for a species, assemblage, or community of wetland birds. While some of these ideas may seem to lack immediate application because of their broad or "generic" level, many relate to wetland habitats regardless of size and location and have implications for a person desiring to develop a small wetland for personal interests or for a team involved in restoration of major areas of global importance. Innovation is of the essence, but conservation approaches must be based on gathering good background information on species and communities typical of the site or area, recognition of patterns or models that may apply when data are minimal, adherence to sound biological and ecological principles, establishment of attainable goals, constant monitoring of progress, and periodic re-design of the strategy.

12.1 Influence of bird biology on conservation approaches

These comments or issues tend to be species- or group-oriented and often are related to territory and home-range patterns and sizes. However, it is essential to recognize that natural or induced changes in the interest of one wetland species or group influence all species within the habitat.

1. Many wetland birds are species of small habitat patches for certain life functions, as wetlands are the units of their spatial world. In some cases, it would seem that such birds may be more easily conserved and their habitats managed because needs can be met in small and dispersed areas. However, we know too little about the loosely knit social behavior of small-patch birds occupying larger regions. Moreover, we cannot ignore other species that only use large wetlands or specific habitats within large areas, requiring that differing scales must be considered for differing species and groups.

2. Functional habitat units for nesting, resting, feeding, roosting, migration stopovers, and wintering all are essential for completion of the annual life cycle and, therefore, must be considered equally important until we measure the relative adaptability the bird has for each. Dependent upon avian mobility, such units could be (i) different sites used at different seasons within a large wetland; (ii) wetland sites located in several different wetlands (of a complex) in the same region or; (iii) wetlands in widely separated geographical areas that are on a global scale.

3. Migration strategies and resource distribution of coastal and interior wetland birds often differ, influencing stopover locations, timing, and duration, as dictated by predictability of water and food conditions.

4. Wetlands near migration corridors (e.g., along rivers and coastlines) may be used by more bird species and numbers than are isolated wetlands – depending upon their size and habitat diversity. However, this may be a function of stopover duration based on feeding strategies, weather conditions, etc. In either case, the duration of use may seem too short to be important, but such habitat is vital as a link in the dynamic life history of a mobile species.

5. Although a bird's habitat needs may seem simple, species' preferences for different vegetation structure, foods, and water depth and dynamics often are poorly known and, therefore impossible to manage for or replicate without extended observations and interpretation based on similar species elsewhere.

6. Where either specific habitat needs or patterns for groups are known, success at creating suitable living conditions has been great, but it can be costly in time and money. Prominent among direct and relatively inexpensive measures are the provision of nest structures that encourage selection in a particular wetland where the species has bred previously. These species include hole-nesting birds like North American Wood Ducks or Prothonotary Warblers, platform-nesting birds like Ospreys, Peregrine Falcons, and Canada Geese, and sandbar birds like Least Terns. Artificial structures are regarded negatively by many, and their use should be minimized to periods before natural structures have developed, such as in restored or created wetlands.

7. Planted food patches and artificial feeding attract field-feeding, seed-eating birds like sparrows, ducks, geese, and cranes but also less desirable groups like blackbirds (Icterids). Direct feeding is of course very costly and only a short-term solution often used to divert birds from agricultural crops. In most cases, natural seed banks provide adequate food when water regimes are properly managed.

8. Management of invertebrate foods is more difficult but becoming better understood and more feasible in wetlands with controlled water regimes. Typically, they are incidental to drawdowns designed to regulate vegetation, but water regulation to create mudflats for shorebirds or reduce water depths for duck feeding is more common (e.g., Helmers 1992).

12.2 Population and species influences on conservation strategies

There is a strong tendency to identify single species of concern or populations of that species because we observe, census, study, and develop a fondness for favored species. Endangered species' programs have spent sizable sums of money (unless you compare them with items in defense budgets) in efforts to save species like Whooping Cranes, and environmentally aware taxpayers seem satisfied that we should. Moreover, there probably was no alternative way to stir the public to an awareness of this need because of the scope and level of understanding of other approaches. In focusing on species, however, we have identified several habitat-related problems and benefits.

Regulations to protect single species may create problems with other endangered or threatened species. For example, managing water regimes in the Florida Everglades to stabilize food resources for Snail Kites tends to reduce food availability for Wood Storks, which are most successful with declining water levels. Where wetland tracts are large and dispersed enough to provide for both species but in different areas, the long-term values of increased wetland productivity that results from the dynamic water regimes and the natural mobility of the birds probably produces the most natural and self-sustaining strategy for both.

Temporal and spatial scales commonly are not considered. A wetland area that serves one migratory or mobile bird species at one season will also serve many others at other times, and resource needs may differ drastically, thus requiring considerable knowledge of avian biology and potential management strategies.

Wetland birds sometimes respond quickly to habitat management, creating the illusion that all wetland birds are easily managed for and populations enhanced at will. In fact, we often are attracting local birds and have no quantitative measure of the population consequences of the action and may not be achieving goals for whole assemblages or communities in jeopardy. Such efforts also may attract species like Red-winged Blackbirds, locally considered as nuisances or agricultural pests.

Although the abundance of many species of wetland birds seem seriously affected by one or several years of drought, their reproductive strategies are products of such evolutionary stresses and most seem to have compensatory mechanisms of mobility to alternative sites, large clutches of eggs during good years, and multiple nesting attempts to ensure a high reproductive rate when conditions allow. Therefore, we should not expect annual constancy in populations when habitats are naturally unstable at various time scales.

12.3 Bird assemblage and community-level issues

The above examples make it clear that we must always consider a proposed conservation strategy from a habitat perspective as related to assemblages, community, or ecosystem. This complicates the method of study, analysis of data, and the ability to tease out interrelationships important in developing a conservation plan, assessing impacts, or evaluating the success of wetland restoration. As a result, we tend to focus on small groups of species chosen because of visibility or index values. However this is done, a multi-species approach broadens our viewpoint and enhances our understanding of the true complexity of what overtly seems like a "simple" system.

Although it is easier to identify, study, and correlate data on a population of a single species with habitat-related causes of declines, interrelationships among species are lost. Therefore, we tend to draw conclusions of values and needs without data on all species that may be affected by the environmental variables for which the same conservation approaches may be useful. Most commonly this favors the conspicuous and highly valued (e.g., a game duck) over the obscure and less-widely valued (e.g., wren, warbler, or sparrow).

Multispecies use of wetlands may make species-oriented conservation strategies risky for other species and, therefore, impractical for any particular habitat unit. As the main focus of any management strategy for communities of wetland birds, it is ideal to maintain basin wetland complexes, representative large wetlands or lakes, and diverse riverine and coastal geomorphic formations that induce self-sustaining wetlands.

12.4 Wetland influences on bird conservation approaches

The ease and cost of maintenance of wetland habitats depends much on their type. For example, basin wetlands are most common and easily visualized and of a size that makes acquisition or management feasible; as a result they are the usual target of conservation efforts. Slope wetlands often are seasonal entities and also of smaller scale with several options for preservation; they also have a high potential impact from human disturbance (e.g., ski slopes, overgrazing or overbrowsing). Fringing wetlands of large lakes and seas and riverine wetlands of large river systems often intersect with many jurisdictional boundaries; the magnitude of the setting requires cooperative efforts often beyond the authority and capacity of many conservation agencies. Several multinational or inter-state cooperative groups have been formed to manage resources of large lakes and rivers on such boundaries, but few efforts of this type have been made toward wetland conservation. Scientifically based and privately funded organizations like the National Audubon Society, the Nature Conservancy, and the World Wildlife Fund have played vital roles in calling the attention of the public to such problems and inducing government cooperation to conserve such resources. Wetlands of flats are so imperceptible as entities that most people do not understand that they are at times wetlands, and these are one of the first areas to be encroached by human activity such as agriculture and urban development. Several major cities of the world are built on drained wetlands and their soils reflect this in decomposition of the organic material and resulting sinking of buildings, roads, and other structures. Moreover, they are regularly and easily flooded. Preservation and restoration of some types of wetland are no longer possible in some areas because the original water regimes that formed the wetland areas may no longer exist.

The concepts of island biography theory have been applied to conservation of birds and preserve design (Diamond 1975) and also have been applied to basin wetlands; however, they are applicable to fringing and riverine wetlands as well because they form linkages between other wetland types. These ideas have influenced much of current thinking on both within and among wetland comparison.

1. Wetlands with great structural and vegetation diversity (both above and within the water column) should have more species and probably numbers of individuals than less diverse units. Although uniform and dense stands may not attract a great diversity of birds, some species favor such habitats and must be considered in conservation actions that result in modification of habitat structure.

2. Open-water wetlands with food resources but without significant standing vegetation will attract nonbreeding (i.e., migrant or wintering) over breeding birds, because of the absence of cover for isolation of pairs, or nests and young. Grebe colonies may be exceptions to this when they build nests of submergent vegetation, or flamingos that build nests of mud on mudflats, but such colonies are very vulnerable to wind damage and flooding. In addition to water depth, emergent vegetation is a product of natural enrichment (eutrophication), often exacerbated by human pollutants; this process may be used advantageously in some management plans if obvious limitations are applied.

3. Open-water wetlands also may be less rich in primary productivity than vegetated areas, may have deeper water, or may have a flooding history that has influenced the vegetation. Birds tend to use these wetlands more during mild weather than in severe and windy weather, and these areas are most attractive to flocks of social species oriented to a single food type which is prominent in that locale, or for roosting.

4. Wetlands of larger size tend to have more bird species than smaller wetlands, typically because of increased habitat diversity, but comparisons used as a measure of success ("index" wetlands) must involve similar wetland types in similar climatic regimes having similar depths and seasonal water regimes. Because of the influence of species, life-history strategies, comparisons are best based on comparable habitats (i.e., vegetation structure, water regimes, depths, etc.) and on birds per unit area.

5. Dynamic and ephemeral habitats (e.g., mudflats, sandbars, and meadows with temporary flooding or drying) are typically undervalued by casual observers but have been shown to be rich in value for food and rest and are crucial stopover sites as well as breeding areas for species specialized to those unique habitats. They may be used only a few weeks or months a year in some areas but are essential for the survival of a global fauna.

6. Wetter is not necessarily better (Weller 1988a). Numerous observations suggest that continuous high water levels in normally dynamic wetlands reduces bird species richness or their suitability for a selected species. Moreover, permanent changes in vegetation and aquatic animals may result, thereby creating a different wetland type. Temporary flooding of formerly dry areas can, however, be very attractive for short periods because of newly available food resources and loafing sites.

7. Extreme periodic drought, while harmful to bird use in the short term, is essential for (i) cycling of nutrients deposited in anaerobic wetland substrates, and (ii) re-vegetation of shallow wetlands that otherwise may remain as open-water or submergent aquatic beds in both saline and freshwater sites. Extended droughts may result in increased salinity in some areas and even salt deposits; these may change the vegetation drastically, resulting in lengthy periods when the area is attractive only to those species of birds especially tolerant of these conditions (e.g., avocets and stilts in salt pans).

8. From a conservation standpoint, severe drought is of greater concern because most wetlands are permanently drained during this event for alternative uses.

9. Bird diversity alone is not the best measure of the value of a wetland; for example, Everglade Kites live in wetlands that are simple, relative to some others, but we would not encourage the elimination of these wetlands in favor of other more complex or species-rich ones.

10. We must evaluate species richness and diversity of assemblages on large spatial and temporal scales. All birds will not be equally abundant at one time or under all wetland conditions, but their dynamics will change over time and area.

12.4.2 AMONG-WETLAND COMPARISONS

1. Bird species diversity of two wetlands of the same size will be greater in the wetland with the greatest habitat diversity.

2. Abundance and probably biomass of birds in two wetlands of the same size will be greater in the wetland with the greatest nutrient riches (e.g., more eutrophic) because the potentially greater productivity is reflected in invertebrates and other community components that may be important resources.

3. The likelihood that two wetlands will have the same bird species composition is greater in wetlands of the same type that are close together rather than in those that are far apart. In mobile species, two wetlands that are close together may be used by the same individuals.

12.5 Values of wetland complexes or clusters

1. A cluster of homogeneous wetlands should be more attractive and productive than an equal number of similar but isolated wetlands because of the flexibility of dispersing individuals that can tap similar resources at different times.

2. A cluster of heterogeneous wetlands should be more attractive and productive than a similar cluster of homogeneous wetlands, because

of the diversity of habitat resources and structure available at any one time.

3. The attractiveness of a group of small wetlands to a single bird species compared with that of solitary wetlands of the same number and size will differ only if the home range of the species incorporates these several wetlands, thereby enhancing dependability and quantity of resources.

4. As a result of water variation over seasons or years, wetland complexes are more likely to have at least some wetlands in a water and plant regime favorable to a particular species, thus ensuring diverse species' representation in the geographical area. Moreover, variation in rainfall patterns will influence the number of wetlands and their water levels, thereby influencing richness and productivity at a given time.

5. Human impacts influence some wetland types more severely than others, resulting in reduced attractiveness and productivity in some years (Krapu *et al.* 1997) or total loss (Weller 1988b).

12.6 Values of wetland water fluctuations

1. Long-term water dynamics dictate wetland type, including vegetation and dependent bird life, and this water variability is essential if these natural features are to be retained (Weller 1988b). Diverting water to increase water stability, or damming a stream to create an impoundment, changes the wetland types radically, and significant responses of birds are inevitable.

2. More stable wetland types may be no more or less desirable than dynamic ones, because they merely differ in physical influences and, therefore, support different wetland bird assemblages at different densities, times, and perhaps for different functions.

3. Manipulations of water regimes are possible especially in basin wetlands, where they are used to enhance and/or direct species composition of the system, but many factors influence the advisability of such actions, including regional wetland plans.

4. Water quality must be maintained through upslope or upstream management to minimize sedimentation, effluents, and pollutants in all wetland types.

5. Maintenance of traditional water quantity and quality for wetlands may be of greater importance than any other conservation action, and effects are especially evident in freshwater inflow to saline waters or flooding regimes in riverine systems.

6. Although more predictable in water regimes than most basin or riverine wetlands, tidally influenced coastal water regimes do vary seasonally and from year to year because of different celestial gravitational forces, wind action, and freshwater inflow.

12.7 Vegetation dynamics

Understanding wetland plant succession by type and region is essential to providing quality habitat for wetland birds to use for specific needs such as cover, nesting sites, and foods. Results of managing plant succession always are influenced by chance but most birds seem sufficiently adaptable in use of foods or cover that the plant species itself is less important than the general type of food produced (e.g., seed size, nutritional quality, and location. Basin configuration and shoreline slope are important both in vegetation management and in bird use because they affect food availability and hence bird foraging tactics. Upslope management for water quality and quantity directly dictates plant communities in the wetland and thereby its attractiveness to birds and other wildlife.

References

Diamond, J. M. (1975). The island dilemma: lessons of modern biogeographic studies on the designs of natural preserves. *Biological Conservation* 7, 129–46.

Helmers, D. L. (1992). *Shorebird Management Manual.* Publication No. 3. Manomet, MA: Wetlands for the Americas.

Krapu, G. L., Greenwood, R. J., Dwyer, C. P., Kraft, K. M., and Cowardin, L. M. (1997). Wetland use, settling patterns, and recruitments in mallards. *Journal of Wildlife Management* 61, 736–46.

Weller, M. W. (1988a). The influence of hydrologic maxima and minima on wildlife habitat and production values of wetlands. In *Wetland Hydrology*, eds. J. Kusler and G. Brooks, pp. 50–60. Technical Report 6. Berne, NY: Association of State Wetland Managers.

Weller, M. W. (1988b). Issues and approaches in assessing cumulative impacts on wildlife habitat in wetlands. *Environmental Management* 12, 695–701.

13

Measures of bird habitat use and quality

To apply some of the concepts and principles noted in previous chapters, biologists typically measure habitat quality for birds for a variety of purposes (Adamus *et al.* 1987, 1991, Bookout 1994): (i) ornithologists or ecologists studying the biology and ecology of individual species or groups; (ii) ecologists studying how birds are distributed in space and time; (iii) wildlife managers measuring habitat use or habitat quality to assess the effectiveness of some form of management; (iv) consultants selecting alternatives for habitat creation or restoration projects; (v) conservation biologists setting priorities for purchase or lease options to protect species, habitats, or communities; (vi) conservation or regulatory agencies evaluating the effectiveness of a wetland mitigation (i.e., replacement) project and subsequent monitoring of success in meeting goals; and (vii) regulatory biologists and population managers following trends in populations, measuring effects of harvest regulations, or evaluating environmental impacts,

Population size and density often are used as indicators of good or poor habitat, but such data have numerous biases and are expensive to obtain. However gathered, it is also essential to have measurements of habitat features to correlate with the population indices to provide some relative scale of quality. Measurement of habitat condition for a particular animal commonly is termed **habitat suitability**, but some workers advise against use of the word "suitability" as redundant (i.e., it either is or is not habitat) and recommend the word "quality" instead (Hall, Krausman and Morrison 1997). However, "suitability" is an integral part of a widely used evaluation system (see below) that describes a range of features attractive to and indicative of habitat quality for the species in question. Habitat **use** infers the presence or relative occupancy rate of birds in a particular habitat as measured by presence and absence, index of abundance, or complete census of sampling areas. **Selection** is sometimes used as a synonym for use but it is better used to describe the process of choosing from available options (Johnson 1980). If one also has data on **availability** of potential habitat or a specific resource such as food or nest sites, **preference** may be expressed as a percentage of that present and feasible to use (Alldrege and Ratti 1986). This measurement is not a simple task because presence of a habitat or food does not necessarily mean that it is available to the bird. For example, a rich supply of fingernail clams could be located in a lake bottom with water too deep for dabbling ducks to reach even with long necks and, therefore, not available except for those clams in very shallow water. They would be within easy reach of diving ducks, however, and could be utilized with some added energy expenditure. Fish that school near the surface might be available to plunge divers like terns or Brown Pelican, but bottom-dwelling fish would be less visible and, therefore, less vulnerable.

Assessing relative quality of habitats is difficult and involves the relationship of features or resources of the habitat to relative use of the species or group under study (e.g., Paquette and Ankney 1996, Swift, Larson and DeGraaf 1984). For the annual breeding cycle, the ultimate measure of quality is **reproductive success**, as measured by number of young reared per adult or female and by the condition or health of birds within a habitat (Vickery, Hunter and Wells 1993). In variable environments like wetlands, such measurements are influenced by many factors and, therefore, must be based on long-term assessments that demonstrate the range of habitat conditions and their influence on use and success. Some options are examined briefly below.

13.1 Species diversity and population assessment

Although our primary objective is to assess habitat, some measure of the species present and their population sizes typically is the first step. There are really few standardized methods for surveying waterbirds because there are so many species, habitats, and seasonal influences that no one technique works for all. Guidance on survey methods and problems can be derived from sections in summary books such as the Wildlife Society's wildlife techniques manuals (Bookout 1994, Schemnitz 1980), species-oriented summaries that focus on applications (Tacha and Braun 1994), and symposia designed to identify and perfect bird census techniques (Ralph and Scott (1981) is excellent although aimed mainly at terrestrial birds). Some important considerations for any technique include: (i) preliminary experimentation to determine feasibility of the method; (ii) determination of known habitat variability over time to determine whether a technique useful under present conditions will work under modified vegetation or water regimes; (iii) simultaneous gathering of supplemental information on habitat features (e.g., vegetation, water depth, distribution of the population) to relate to census data; (iv) standardization of techniques over time and space where possible to ensure efficacy for breeding versus wintering where needed; (v) evaluation of the technique through the use of repeated surveys (see Colwell and Cooper 1993), with marked birds where feasible; and (vi) trial manipulation of mock data statistically to assess the real potential of deriving significant differences under the conditions of study. Realistically, few of these options are possible because of limits in time and money, and thus we optimize techniques accordingly.

Population surveys or estimates taken by **direct counts** of a single species of bird is the most common parameter used to assess habitat values of an area because this allows us to calculate **density** in areas of known size and features. Several authors have warned of the biases of overemphasizing numbers (van Horne 1983); moreover, the need for such counts really is dependent on the question asked. **Total counts** often are impractical on large areas, and counts on **sample plots** are a necessary substitute; these may be fixed quadrats, line or belt transects, or plots of variable sizes and locations, but the system used depends on the characteristics of the bird species and its amenability to the approach. Randomly selected samples are considered essential by statisticians but we often are forced for logistical reasons to sample what we can reach or see, and our data represent an index to the population of only the sampled area rather that the entire habitat or population that we might like to assess. Another index that can be related to habitat features of sample plots is data on **presence** or **absence** of a species or group of interest. Where required by the behavior of the species in question, **indirect counts** of calls have been useful, especially coordinated with tape recorder playbacks for rails and other inconspicuous birds (Gibbs and Melvin 1993, Glahn 1974, Tacha and Braun 1994). Numbers of birds trapped, trapping success, or mark-recapture data also have been useful indices in some intensive studies. Counts of nests, total young, or young per nest or per female are used for several species such as colonial waterbirds, where nests are conspicuous and can be counted and even photographed from a plane.

Although some of these approaches seem simple enough, all are biased by visibility; wetland birds often are hidden from view especially during nesting and rearing periods, and therefore, visibility varies with seasons, vegetation, and bird behavior. On the positive side, this is the period when many birds call to proclaim territories and communicate with young.

Population data or other information must be compared within similar time and space categories because use patterns of birds are so dynamic. Population data also should be related to available habitat for various functions as indicated by nest sites, feeding areas, and even food resources in cases where one is identifying causal factors. Although it seems obvious, counts of single species should be compared only with data on that species for comparisons of different times, wetlands, or seasons. There has been a tendency to pool closely related species as if they were one to increase statistical validity, but this masks information and reduces biological validity. For example, an increase in numbers of a herbivorous duck like the Gadwall is not comparable with a decrease in an omnivore like the Mallard or Blue-winged Teal; it suggests entirely different food abundance and wetland succession patterns. Grouping all species of a taxonomic group such as waterfowl (ducks, geese, and swans) fuses data on species that represent three or more trophic niches, many habitats, and countless feeding tactics dictated by the wetland and temporal variations in habitat use. However, such approaches may provide data useful for comparison with other high-order groups in multispecies analyses, such as wading or walking birds like herons, which feed on different foods in different ways.

13.1.2 MULTISPECIES LEVEL ASSESSMENT

Community-level assessments usually are based on population data by species recorded on surveys, as mentioned above, from a particular habitat type. By expressing the percentage that each species constitutes of the total, a composite view termed **species composition** is derived. This is an old method that has regained popularity because it reflects diversity of species and relative numbers with little mathematical bias (James and Rathbun 1981). Perhaps the most simple measure of the variety of species present in a specific habitat or area is a species list on which the number of species (**species richness**) is used for comparisons between areas (Pielou 1975). It often is correlated with habitat features and food resources, but without additional and often detailed information on species or habitats, it can mask major habitat differences. For example, one could have 10 species of shorebirds using the invertebrates in one wetland that was mostly mudflat and 10 species of ducks and other swimming waterbirds using different foliage and seeds in another wetland of more permanent water. These are not equivalent and will not provide valuable data unless the analysis addresses the question asked.

The **species diversity index** is a mathematical approach developed as an index to diversity of plants that has been very popular among animal ecologists as well. It includes a measure of both the diversity of bird species (richness above) and relative abundance of each species (evenness). If only this index is published or used in statistical analyses, species identity is lost as is information on resource use that can be inferred from knowing the natural history of those birds.

13.1.3 EXAMPLES OF SOME COMMON BIRD SURVEYS

Because of their economic and recreational importance, or because of threatened status, waterbirds have figured prominently in the development of census techniques for all birds. Moreover, volunteers have played a major role in providing information to aid in bird conservation and population management. Nationwide teams of observers were organized by Wells W. Cooke of the US Biological Survey to make counts and observations on timing of migration in shorebirds and waterfowl in the late 1800s and early 1900s (e.g., Cooke 1906). These formed the basis for understanding many patterns of distribution on a continent-wide basis and for concepts of bird flyways (Lincoln 1935) and corridors (Bellrose 1972).

Measures of bird habitat use and quality

Cooke's work was influential in the design of the National Audubon Society's Annual **Christmas Bird Count**, now nearly 100–years old (National Audubon Society 1997). Wetland birds are included among the total of species counted and the count offers some data not otherwise available. These data vary greatly in coverage and detail because of the variability in number and experience of observers, and in the climatic conditions during the survey, but they imply population trends of species of concern.

A habitat-based survey for birds is the **Breeding Bird Count (BBC)** sponsored by the US Fish and Wildlife Service. Birds are tallied by sightings and calls in a prescribed site and size that then form annual censuses during the breeding period (Robbins, Bystrak and Geissler 1986). This provides patterns of distribution and estimated density for many common species (Price, Droege and Price 1995), and is useful for the analysis of population trends (Sauer and Droege 1990).

Because of their importance for hunting, waterfowl population data have been obtained annually in North America by federal, state, and provincial wildlife agencies for management purposes since the early 1930s, and these valuable data sets can be used for assessment of population changes and relative abundance of species over time. Waterfowl numbers at first were assessed by ground counts of concentrations, but the need for more extensive data in areas unreachable by car, foot, or even canoe resulted in a move to the use of aerial surveys on both breeding and wintering areas (Cowardin and Blohm 1992). Although accuracy is variable by biotic region, standardized methodology in the North American **Waterfowl Breeding Population and Habitat Survey** results in a good annual index to changes in numbers of ponds and breeding pairs in a vast area of the upper Midwest and Canada (Fig. 13.1a) and gives some information on water conditions (Caithhamer and Dubovsky 1997). These comparative surveys help to identify species that do (Fig. 13.1b) and do not (Fig. 13.1c) respond to changed wetland numbers. Surveys after the breeding season provide a **production index** of broods-per-female per water area by regions and form the basis for estimating the fall-flight of dominant species. Wintering survey methods are less standardized but are appropriate to regional populations and habitats; they provide a **winter population index**, which has been used to record changes in distribution but also to identify problems with habitat resources (Baldasarre and Bolen 1994). However, to understand habitat features that cause population changes, information on resource availability and distribution would be essential. A similar winter count focused on nongame birds was initiated by the US Fish and Wildlife Service staff in 1948 and has been continued to date and published annually in the *Journal of Field Ornithology* (e.g., Lowe 1995). This survey is a product of repeated counts on fairly uniform plots, so data on various wetland types can be derived. Similar surveys have been established for various groups around the world, many of which have been in effect for 25 to 50 years or more (e.g., Atkinson-Wiles 1963, Milstein 1968).

(a)

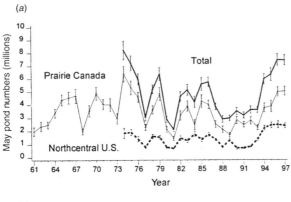

(b)

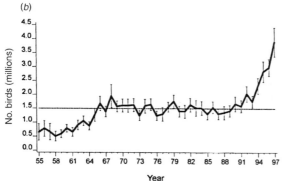

(c)

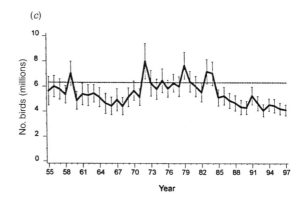

Reduced populations of egrets, herons, and other waders have been cause for concern, and a variety of techniques have been developed and used to assess population changes as well as shifts in distribution. Traditionally, nesting colonies have attracted the most attention and best records (e.g., the *Colonial Bird Register*). Many colonies of nesting birds are in conspicuous sites in trees or shrubs, and annual or periodic surveys have occurred to assess long-term trends (Lange 1992, Nesbitt *et al.* 1982, Portnoy 1978, Runde *et al.* 1991). Broad-scale but systematic censuses by plane and on ground have been used to provide data on colony location, feeding areas, species richness, and general population size. In colonies where thousands of birds nest and are at several levels in the vegetation and, therefore, cannot be counted precisely from the air, ground strip-transects through colonies have been used to estimate numbers of nests (King 1978). Comparisons of helicopter and fixed-wing aircraft for colony surveys show that helicopters are much more accurate (<20% error for white species but much greater for dark birds) than fixed-wing flights (>35% even for white birds)(Kushlan 1979). However, use of systematic, transect-based aerial surveys reportedly underestimates the true wader population by 15 to 20%, which is comparable to the variation for waterfowl surveys. Comparison of aerial and water-level surveys (via airboat) also demonstrates an advantage for colony discovery by aerial methods, espe-

Figure 13.1. Long-term variation in (*a*) May pond numbers based on waterfowl breeding pair surveys in the northern United States and Canada, with population data for Gadwall (*b*) and Lesser Scaup (*c*) that show different responses to water availability (Caithammer and Dubovsky 1997).

cially for white birds like egrets as opposed to dark herons or ibises (Frederick *et al.* 1996).

Numerous individuals and groups worldwide have established long-term surveys of species culminating in detailed avian atlases, and more are now assessing birds by habitat as well as in geographical units. A state-level survey established in North Dakota formed the basis for a habitat-oriented bird book that provided species composition and species density estimates for some groups (Stewart 1975). Re-visits to these sites have resulted in comparative trends for many species and reflect long-term declines for wetland and grassland species (Igl and Johnson 1997). Some waterfowl surveys at the national level show migration routes as well as population density estimates along these routes (Bellrose 1980).

International migrants have been much more difficult to trace, but recent efforts of several shorebird study groups have resulted in the *International Shorebird Survey* and many subgroups have a long history of effort for smaller areas. Numerically, these bird groups represent one of the greatest challenges because of numbers running into the millions at some sites like the Bay of Fundy and areas of similar tidal extremes in Tierra del Fuego, with a complex mix of species as well.

At the other extreme are small-scale but habitat-specific censuses that have been done by boat along lakeshores to associate wader species with specific habitats (Esler 1992), or along coastal islands for species and population counts, but such access is not always available in backwater parts of riverine systems. Counts from boats often are less disturbing than ground surveys and can be repeated regularly with little apparent population effect. Ground counts of nests have been used for some waders, especially those that are solitary or less visible; belt transects often are useful to minimize the labor but assure coverage of various vegetation types (Weller and Frederickson 1974). Radio-tracking of individual birds in flocks using Global Positioning Systems now allows pinpointing of microsites, temporal data on individuals, and meaningful estimates of turnover time, which is essential to estimate real numbers of birds using a site.

The above very incomplete arrays of techniques and issues may be both complicating and discouraging for those who are searching for simple techniques to answer important questions; unfortunately, such techniques do not exist. Choice of technique is dependent upon the questions asked, and the time and money available to gather the information at a satisfactory level. Some census techniques have been refined for 50 years or more and are little more accurate than when first developed, but they now form important indices. Each study is different, and the suggestions I outlined earlier can help to avoid useless data. Statistical requirements complicate data gathering even more and may result in a failure to address the specific question asked. Knowledge of the bird life histories and habitat requirements are essential first steps for ensuring good questions, the development of practical techniques, and thereby useful

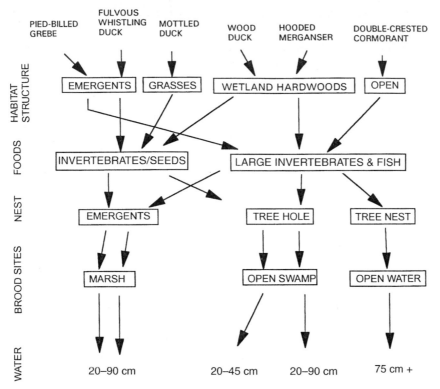

Figure 13.2. Habitat resource influences on swimming water-birds in the Kissimmee River wetlands in relation to foods and habitats selected (Weller (1995) Use of two waterbird guilds as evaluation tools for the Kissimmee River restoration. *Restoration Ecology* 3, 211–14. Reprinted by permission of Blackwell Science, Inc.).

answers; however these answers are not always at the level of detail needed or have the accuracy desired. Whatever is done, measurement of population and species diversity without information on habitats and their status will weaken the understanding of the community in question and the opportunity to apply such population information to conservation assessments.

13.2 Habitat diversity and other wetland quality issues

There are many parameters of habitat that have been correlated statistically with the presence or abundance of species, but insight into the bird's natural history and resource requirements are essential to select the best indicators to measure. The importance of vegetation structure and other physical structures were discussed above. Bird species richness and the species-diversity index generally are directly correlated with habitat diversity of both biological and structural features. The greater the variety of habitats, regardless of the cause, the more likely it is that additional bird species can find suitable habitat (Figs. 13.2, 13.3). However, measuring this diversity is complex because it includes plant diversity and structure, ideally plant and animal food diversity, water characteristics such as depth, quality, and permanence, and many other features. There is no simple system for accurately and quickly measuring all of

Figure 13.3. Habitat patterns and water regimes attractive to egrets and other waders, and typical foods in Kissimmee River wetlands (Weller (1995) Use of two waterbird guilds as evaluation tools for the Kissimmee River restoration. *Restoration Ecology* 3, 211–14. Reprinted by permission of Blackwell Science, Inc.).

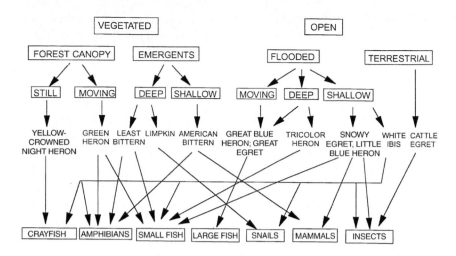

these features, and it is especially frustrating that one can measure many things and still not find the suitable correlate! Moreover, what is important in one season may be meaningless in another.

Horizontal distribution of vegetation and other structural features is the most simple and currently easiest parameter to identify because of the availability of aerial photographs or satellite imagery, computer scanning, and Geographic Information Systems, but these measurements may not demonstrate the expected correlation. Typically, one seeks to measure heterogeneity in horizontal or spatial patterns of cover–water, plant diversity, etc. that influences habitat structure. Size of used habitats (**patch size**) and their **distribution** or **dispersion** are important components used as correlates for species or community distribution and abundance. These data have been most often used in calculating cover–water or cover–cover ratios, amount of edge, and other structural indices that can be correlated with species' presence and density (Beecher 1942, Weller and Spatcher 1965). These do not necessarily indicate the cause of the relationship but do lead one closer to finding it.

Measurements and analyses of data on vertical structure of habitat diversity are more difficult. These usually have compared things like nest height in relation to plant structure (e.g., height, stem thickness and density, foods, water, etc.) compared with that in unused plants. Rarely have these analyses in wetlands incorporated the features of the water column or substrate characteristics.

Habitat patterns derived from a few parameters of simple systems like urban and forest-edge habitats (DeGraaf 1986) combined with computer-based models of pattern recognition have been used to assess and predict the presence and relative abundance of selected bird species for forest birds (Grubb 1988, Williams, Russel and Seitz 1977). Such patterns have not been fully exploited for wetland birds except in relation to a diversity index (Adamus 1995), but the generalized habitat patterns used earlier to describe wetland birds have potential for such conceptual and perhaps mathematical models. Features other than structure have been used to explain the presence and density of selected species, such as food (Chapters 4 and 5), water depth, and water quality (Utschick 1976).

The **Habitat Evaluation Procedure (HEP)** is used by US federal agencies such as the Fish and Wildlife Service in assessing wetland habitat loss or damage. The relative quality of the habitat is based on a simple model called the **Habitat Suitability Index (HSI)**, which uses a few parameters that seem to correlate with a species' presence, abundance, and reproductive success. Such indices have been created for wetland birds as well as for fish, mammals, and invertebrates; these are index ratios with a maximal value of 1.0. These quality estimates then can be used to specify the amount of replacement habitat to compensate for lost habitat but based on quality. If, for example, 100 acres of wetland were to be destroyed by a highway development project which had an HSI of 0.85 for a certain species (or pooled ratio for several key species), $100 \times 0.85 = 85$ **habitat units (HU)** of HSI 1.0 value would be required to replace this habitat. If lower quality habitat (HSI) is used for mitigation (HSI = 0.25), it would require more acreage to replace that lost: 85 HU/0.25 HSI = 340 acres. One can also use this system to document changes over time, or compare two communities, but the system has been used more for single species than for groups.

An analysis based upon the incidence of bird species derived by surveys which are then clustered according to the type and manner in which they use resources (i.e., guilds) stresses the importance of habitat structural diversity, food resources, and water regimes (Short and Burnham 1982). Life-history data of members of an assemblage or community are analyzed as a **guild matrix** according to a series of habitat features viewed as resources for some function like nesting plotted against another resource use such as feeding. Occupied blocks in the matrix reflect species using those guilds, allowing a visual assessment of the complexity of a community based on the diversity of resources. The resulting guild matrix also is a very useful approach to documenting changes over time resulting from either natural events or human impacts and results in a conceptual model of the habitat pattern that allows prediction of species composition on the basis of habitat features (Short 1989) (Fig. 13.4).

Several broader and less bird-oriented techniques have been developed and used by various agencies to assess the overall ecosystem or community components, at least on a relative scale. The system developed for the US Federal Highway Administration (Adamus *et al.* 1987, 1991) is among the best because it qualitatively judges 75 or so parameters of the wetland and its components; it uses waterbirds in particular because they indirectly reflect important habitat quality features that are important to wetland functions other than wildlife habitat (e.g., water purification, sediment trapping, erosion control, fish habitat). The Adamus evaluation system is hierarchical, allowing the user to incorporate any level of detail desirable, based on available knowledge or for which funds are available.

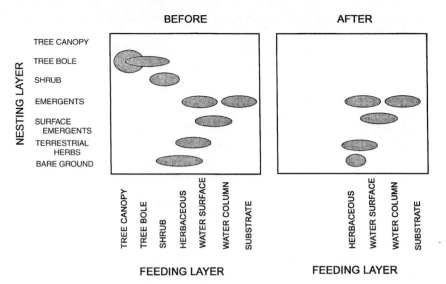

BEFORE AFTER

NESTING LAYER

TREE CANOPY
TREE BOLE
SHRUB
EMERGENTS
SURFACE EMERGENTS
TERRESTRIAL HERBS
BARE GROUND

TREE CANOPY
TREE BOLE
SHRUB
HERBACEOUS
WATER SURFACE
WATER COLUMN
SUBSTRATE

FEEDING LAYER

HERBACEOUS
WATER SURFACE
WATER COLUMN
SUBSTRATE

FEEDING LAYER

Figure 13.4. A guild matrix analysis structured by layers of vegetation used for feeding and nesting with hypothetical distribution of birds before and after the elimination of trees and shrubs in a wetland (based on concepts and forms suggested by Short (1989)).

New approaches are constantly being developed, often to reduce time and costs, but it should be apparent that these will fail if not based on a thorough understanding of the manner in which various bird species use such habitats for food or nest sites, etc. and of the influence of habitat diversity and dynamics on populations and species composition. From a practical standpoint, it may be impossible to gather all the desired information, but one often can exploit existing biological data, perhaps gathered for other purposes, that may be useful in conservation and other ecological applications.

References

Adamus, P. R. (1995). Validating a habitat evaluation method for predicting avian richness. *Wildlife Society Bulletin* 23, 743–9.

Adamus, P. R., Clairain E. J., Jr, Smith, R. D., and Young, R. E. (1987). *Wetland Evaluation Technique (WET)*; Vol. II, Technical Report Y-87. Vicksburg, MS: U S Army Corps of Engineers, Waterways Experiment Station.

Adamus, P. R., Stockwell, L. T., Clairain, E. J. Jr, Morrow, M. E., Rozas, L. P., and Smith, R. D. (1991). *Wetland Evaluation Technique (WET)*, Vol. I. Technical Report WRP-DE-2. Vicksburg, MS: U S Army Corps of Engineers, Waterways Experiment Station.

Alldrege, J. R. and Ratti, J. T. (1986). Comparison of some statistical techniques for analysis of resource selection. *Journal of Wildlife Management* 50, 157–65.

Atkinson-Wiles, G. L. (1963). *Wildfowl in Great Britain.* Nature Conservancy Monograph No. 3. London: Her Majesty's Stationary Office.

Baldassare, G. A. and Bolen, E. G. (1994). *Waterfowl ecology and management.* New York: Wiley.

Beecher, W. J. (1942). *Nesting birds and the vegetative substrate.* Chicago, IL: Chicago Ornithological Society.

Bellrose, F. C. (1972). Mallard migration corridors as revealed by population distribu-
    tion, banding and radar. In *Population ecology of migratory birds*, pp. 3–26.
    Wildlife Research Report 2. Washington, DC: U S Fish and Wildlife Service.

Bellrose, F. C. (1980). *Ducks, geese and swans of North America*, 3rd edn. Washington,
    DC: Wildlife Management Institute.

Bookout, T. A.(ed.) (1994). *Research and management for wildlife and habitats* 5th
    edn. Bethesda, MD: The Wildlife Society.

Caithamer, D. F. and Dubovsky, J. A. (1997). *Waterfowl population status*, 1997.
    Washington, DC: U S Fish and Wildlife Service.

Colwell, M. A. and Cooper. R. J. (1993). Estimates of coastal shorebird abundance:
    the importance of multiple counts. *Journal of Field Ornithology* 64, 293–301.

Cooke, W. W. (1906). *Distribution and migration of North American ducks, geese and
    swans.* Biological Survey Bulletin No. 26. Washington, DC: U S Department of
    Agriculture.

Cowardin, L. M. and Blohm, R. J. (1992). Breeding population inventories and mea-
    sures of recruitment. In *Ecology and management of breeding waterfowl*, eds.
    B. D. J. Batt, A. D. Afton, M. G. Anderson, C. D. Ankney, D. H. Johnson, J. A.
    Kadlec, and G. L. Krapu, pp.423–45. Minneapolis, MN: University of Minnesota
    Press.

DeGraaf, R. M. (1986). Urban bird habitat relationships: application to landscape
    design. *Transactions of the North American Wildlife & Natural Resources
    Conference* 51, 232–48.

Esler, D. (1992). Habitat use by piscivorous birds in a power plant cooling reservoir.
    Journal of Field Ornithology 63, 241–9.

Frederick, P. C., Towles, T., Sawicki, R. J., and Bancroft, G. T. (1996). Comparison
    of aerial and ground techniques for discovery and census of wading bird
    (Ciconiiformes) nesting colonies. *Condor* 98, 837–41.

Gibbs, J. P. and Melvin, S. M. (1993). Call-response surveys for monitoring breeding
    waterbirds. *Journal of Wildlife Management* 57, 27–34.

Glahn, J. F. (1974). Study of breeding rails with recorded calls in North-Central
    Colorado. *Wilson Bulletin* 86, 206–14.

Grubb, T. G. (1988). Pattern recognition – a simple model for evaluating wildlife
    habitat. *US Forest Service Rocky Mountain Experiment Station Research Note*
    RM-487.

Hall, L. S., Krausman, P. R., and Morrison, M. L. (1997). The habitat concept and a
    plea for standard terminology. *Wildlife Society Bulletin* 25, 173–82.

Igl, L. D. and Johnson, D. H. (1997). Changes in breeding bird populations in North
    Dakota: 1967–1992–3. *Auk* 114, 74–92.

James, F. C. and Rathbun, S. (1981). Rarefaction, relative abundance, and diversity of
    avian communities. *Auk* 98, 785–800.

Johnson, D. H. (1980). The comparison of usage and availability measurements for
    evaluating resource preference. *Ecology* 61, 65–71.

King, K. A. (1978). Colonial wading bird survey and census techniques. In *Wading
    birds*, eds. A. C. Sprunt IV, J. C. Ogden, and S. Winckler, pp. 155–9. Research
    Report 7. New York: National Audubon Society.

Kushlan, J. A. (1979). Effects of helicopter censuses on wading bird colonies. *Journal
    of Wildlife Management* 43, 756–60.

Lange, M. L. (1992). *Texas coastal waterbird colonies, 1973–1990*. Angleton, TX: U S Fish and Wildlife Service.

Lincoln, F. C. (1935). The waterfowl flyways of North America. *US Department of Agriculture Circular 342*, 1–12.

Lowe, J. D. (1995). Resident and winter bird population status of 1994. *Journal of Field Ornithology* Supplement **66**, 1–129.

Milstein, P. L. S. (1968). Practical census methods for waterfowl conservation in the Transvaal, South Africa. *Fauna and Flora* **19**, 45–50.

National Audubon Society (1997). The ninety-fifth Christmas bird count. *National Audubon Society Field Notes* **52**, 135–710.

Nesbitt, S. A., Ogden, J. C., Kale II, H. W., Patty, B. W. and Rowse, L. A. (1982). *Florida atlas of breeding sites for herons and their allies 1976–78*. FWS/OBS 81/49. Washington, DC: U S Fish and Wildlife Service Office of Biological Services.

Paquette, G. A. and Ankney, C. D. (1996). Wetland selection by American Green-winged Teal in British Columbia. *Condor* **98**, 27–37.

Pielou, E. C. (1975). *Ecological diversity*. New York: Wiley.

Portnoy, J. W. (1978). A wading bird inventory of coastal Louisiana. In *Wading birds*, eds. A. C. Sprunt IV, J. C. Ogden, and S. Winckler, pp.227–34. Research Report 7. New York: National Audubon Society.

Price, J., Droege, S., and Price, A. (1995). *The summer atlas of North American birds*. San Diego, CA: Academic Press.

Ralph, C. J. and Scott, J. M. (eds.). (1981). Estimating numbers of terrestrial birds. *Studies in Avian Biology* **6**, 1–630.

Robbins, C. S., Bystrak, D., and Geissler, P. H. (1986). *The breeding bird survey: Its first fifteen years 1965–1979*. Resource Publication No. 157. Washington DC: US Fish & Wildlife Service.

Runde, D E., Gore, J. A., Hovis, J. A., Robson, M S., and Southall, P. D. (1991). *Florida atlas of breeding bird sites for herons and their allies*. Nongame Wildlife Program Technical Report 10. Tallahasse, FL: Florida Game & Fresh Water Fish Commission.

Sauer, J. R. and Droege S. (eds.) (1990). *Survey design and statistical methods for the estimation of avian population trends*. Washington, DC: US Fish & Wildlife Service. Biological Report No. **90**.

Schemnitz, S. D. (1980). *Wildlife management techniques manual*, 4th edn. Washington, DC: The Wildlife Society.

Short, H. L. (1989). A wildlife habitat model for predicting effects of human activities on nesting birds. In *Freshwater wetlands and wildlife*, eds. R. R. Sharitz and J. W. Gibbons, pp. 957–73. Washington, DC: U S Department of Energy.

Short, H. L. and Burnham, K. P. (1982). *Technique for structuring wildlife guilds to evaluate impacts on wildlife communities*. Special Scientific Report – Wildlife 44. Washington, DC: US Fish and Wildlife Service.

Stewart, R. E. (1975). *Breeding birds of North Dakota*. Fargo, ND: Tri-College Center for Environmental Studies.

Swift, B. L., Larson, J. S., and DeGraaf, R. M. (1984). Relationship of breeding bird density and diversity to habitat variables in forested wetlands. *Wilson Bulletin* **96**, 48–59.

Tacha, T. and Braun, C. (1994). *Migratory shore and upland game bird management in North America*. Washington, DC: International Fish and Wildlife Agencies.

Utschick, H. (1976). Die Wasservogel als Indikatoren für den okologischen Zustand von Seen (with English summary). *Verhandlungen der Ornithologischen Gesellschaft in Bayern* **22**(3/4), 395–438.

van Horne, B. (1983). Density as a misleading indicator of habitat quality. *Journal of Wildlife Management* **47**, 893–901.

Vickery, P. D., Hunter, M. L. Jr, and Wells, J. V. (1993). Use of a new reproductive index to evaluate relationship between habitat quality and breeding success. *Auk* **109**, 697–705.

Weller, M. W. (1995). Use of two waterbird guilds as evaluation tools for the Kissimmee River restoration. *Restoration Ecology* **3**, 211–24.

Weller, M. W. and Fredrickson, L. H. (1974). Avian ecology of a managed glacial marsh. *Living Bird* **12**, 269–91.

Weller, M. W. and Spatcher, C. E. (1965). Role of habitat in the distribution and abundance of marsh birds. *Iowa State University Agriculture & Home Economics Experiment Station Special Report* **43**, 1–31.

Williams, G. L., Russell, K. R. and Seitz, W. (1977). Pattern recognition as a tool in the ecological analysis of habitat. In *Classification inventory, and analysis of fish and wildlife habitat: proceedings of a national symposium; Phoenix, AZ*, pp. 521–31. Washington, DC: U S Fish & Wildlife Service.

Further reading

Bellrose, F. C. (1968). Waterfowl migration corridors. *Illinois Natural History Survey Biological Notes* **61**, 1–24.

Cooke, W. W. (1914). *Distribution and migration of North American rails and their allies.* Biological Survey Bulletin No. **128**. Washington, DC: US Department of Agriculture.

Furness, R. W. and Greenwood, J. J. D. (eds.) (1993). *Birds as monitors of environmental change.* London: Chapman & Hall.

Johnson, D. H. and Grier, J. W. (1988). Determinants of breeding distribution of ducks. *Wildlife Monographs* **100**, 1–37.

Maurer, B. A. (1986). Predicting habitat quality for grassland birds using density–habitat correlations. *Journal of Wildlife Management* **50**, 556–66.

Mikol, S. A. (1980). *Field guidelines for using transects to sample nongame bird populations.* Publication 80/58. Washington, DC: U S Fish & Wildlife Service Office of Biological Services.

Reichholf, J. H. (1976). The possible use of the aquatic bird communities as indicators for the ecological conditions of wetlands. *Landschaftsrerband Stadt* **8**, 125–9.

Root, R. B. (1967). The niche exploitation pattern of the blue–gray gnatcatcher. *Ecological Monographs* **37**, 317–50.

Schamberger, M. and Farmer, A. (1978). The habitat evaluation procedures: their application in project planning and impact evaluation. *Transactions of the North American Wildlife and Natural Resources Conference* **43**, 274–83.

Weller, M. W. (1988). Issues and approaches in assessing cumulative impacts on wildlife habitat in wetlands. *Environmental Management* **12**, 695–701.

14

Current status and some conservation problems

Population declines of numerous migrant songbirds have rightfully aroused great public concern about the recent loss of tropical forest habitats. However, extensive coastlines and vast grasslands of North and South America, Africa, and Australia have been under human impact for hundreds of years. Migratory wetland birds of the northern hemisphere that skirted or flew over tropical forests also must have been greatly affected by coastal urbanization and development and by drainage of freshwater wetlands for agriculture, especially since the late 1800s. Such wetland losses undoubtedly influenced the decline of the northern water and grassland birds much as we now see in tropical forest regions. Scientists working with shorebirds and waterfowl have recognized and promoted multinational and global-scale conservation programs (Myers *et al.* 1987); such goals will continue to be one of our greatest conservation challenges.

Conservation of wetland birds resembles that of other groups of animal because all require a major emphasis on preservation of habitat, but it differs in its landscape or global scale. To this point, we have been discussing how habitat resources influence the success of wetland birds, and even how they compete for resources among themselves; we now examine how they compete with society for resources at several levels. Obvious, there is a tremendous geographic variation, and emphasis here will be mostly on the American scene because I know that best – not that it is necessarily better. In fact, most societies seem to have profited little from the past mistakes of other nations, even when communication is good.

14.1 Wetland losses

To appreciate the current status of wetland habitats for birds, we need to examine briefly how society has dealt with wetland habitats in most of the world. A number of early cultures learned to live with and in wetlands for their resources, for example the marsh Arabs of Iraq, where massive living quarters once were built from reeds (Maxwell 1957); the Uro Indians of Lake Titicaca in Peru, who built small houses on floating mats and constructed their famous canoes of bulrushes; the boat dwellers of Pakistan marshes who fish and also use mounted waterbirds on their heads to hide their underwater approach to living birds, much as the ancient Egyptians did; the Calusa Indians of the Everglades (Douglas 1947) and other coastal tribes of the Gulf of Mexico, who used clams, fish, and water plants as their main food; and the tribes of the once-vast Stillwater Marsh and Carson Sink in the Great Basin of the western USA (Larson and Kelly 1995), well-known for their beautiful duck decoys made of emergent marsh plants like California bulrush or tule (Earnest 1965). Many river societies of South America, Asia, and Africa have devised mobile life-styles dictated by flood waters, and housing designs to minimize loss or damage. All societies have used vertebrate resources of wetlands and some still do: hunting, trapping, fishing, clamming, using ducks domesticated for egg or meat production, and some have used birds as "watchdogs" (Southern Screamer of South America). Fish-dependent cultures like those of the Orient and Middle East have for centuries used trained cormorants or herons to capture food rather than using the less dependable arrow, spear, or fish-hook. These reflect the early origins of humans as functional entities in natural community relationships, an example of which is generalized in a food web of the Kafue River flats of Zambia (Fig. 14.1) (Maltby 1986).

Most well-developed societies have tried to eliminate wetlands through drainage ditches and have modified water flow via dams or diversion structures for reasons of convenience, disease control, and re-direction of the water for use in irrigated agriculture, urbanization, and industry (Dugan 1990). Wetland loss worldwide has been calamitous – but it is poorly documented in most countries. Considering that wetlands constituted only 5–7% of the land area in most regions of the contiguous USA, why are wetlands so in demand and why are we so often in conflict with such natural land features? And why, after eliminating about 50% of the original wetlands, are we still exerting pressure to change or eliminate wetland? Moreover, some parts of the world exceed this figure, having lost 80 to 98% (Dahl 1990, Dahl, Johnson and Frayer 1991, Maltby 1986, Tiner 1984). In terms of major bird habitats, only grasslands have suffered greater loss in area or quality in North America and probably around the world.

Pressures on wetlands varies by region, land-use, and human perspective. In agricultural areas around the world, it is often an economic need by individuals and families to maximize earnings from every acre of available land, but these efforts vary in their ecological and economic efficiency. In major urban areas, losses result from the association of developments with water: rivers, lakes, and coastlines. As cities grow, choice land for continued development is no longer available, areas considered of "low-value" are used for commercial or industrial developments or built-up for housing by filling. Even if maintained as wet areas by necessity, they often are modified by impoundments for water storage or dedicated as intensively managed city and county parks, no longer performing the multiple functions of dynamic natural systems. Therefore,

Current status: conservation

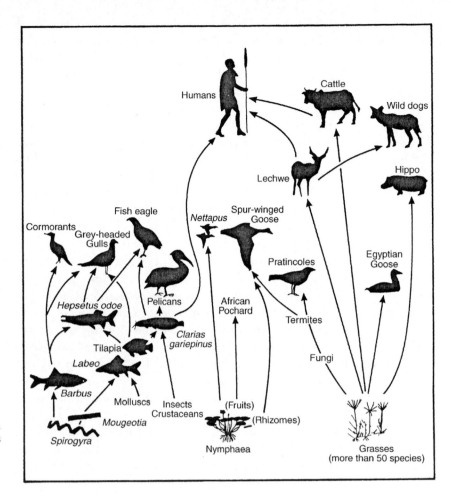

Figure 14.1. Relationships between primary production of hydrophytes and waterbirds, grazing ungulates, and humans in the Kafue River floodplain, Zambia (Maltby 1986, with permission).

many of our urban and suburban nature preserves are relics of a once common habitat type.

Appreciation of wetland values has come slowly to most developed nations, initially as part of the conservation movement and more recently through recognition of the ecological and economic values of wetland ecosystems (Good, Whigham and Simpson 1979, Gosselink, Odum and Pope 1973, Greeson, Clark and Clark 1978, Maltby 1986). Interest in preservation of unique lands in North America arose during the late 1800s and early 1900s after recognition and concern over the loss of rare species. Much of the motivation was prompted by loss of large mammals and terrestrial birds, for example Passenger Pigeons, which concerned hunters. However, declines of waterbirds such as Eskimo Curlew, Great Auk, Labrador Duck, Trumpeter Swan, Whooping Crane (Forbush 1916) and the Ivory-billed Woodpecker became of special interest with the potential damage threatened by passage of the United States Reclamation Act in 1902. This Act was directed toward a national survey of "watery wastelands" that could be re-claimed, mainly to encourage drainage, regulation, and impounding of streams, and for water use in irrigation. One secondary result was an information base on the extent of wetlands in that period, from which we now estimate losses. Similar actions have occurred around the world, often much earlier than in North America.

Drainage of larger and deeper wetlands has required major engineering efforts, suitable topography, or pump technology. Yes, some wetland drainage schemes actually have failed because people tried to drain them uphill! Much of this technology was developed in the Netherlands and the United Kingdom for land reclamation or in Asia for rice production but has been transported around the world. In many cases, drainage water has been moved via rivers and into the sea, but in more inland settings, drainage typically flows from smaller areas to large collecting basins, lakes, or sumps. Therefore, enlargement of wetlands or stabilization of lakes as reservoirs have been the products of drainage or irrigation programs, all enhanced by the development of huge earth-moving equipment allowing massive land changes heretofore impossible. Losses also have been facilitated by our failure to develop national land management goals that recognize the long-term wetland values of flood control and aquatic resources compared with immediate or short-range economic benefits. There are numerous examples of how drainage systems have worsened riverine and urban flooding regimes and have been destructive to wetlands (Miller and Nudds 1996), and how their present maintenance continues to have an impact on wetland quality and predictability with significant effects on waterbirds (Krapu 1996).

The Florida Everglades, perhaps the largest wetland in North America, is a product of rivers like the Kissimmee or a river-like sheetwater flow from north to south. It has been modified by impoundments for water supply or alternative uses, and the northern segment now is a series of three large impoundments that protect nearby developments from flooding and ensure urban water supplies and agricultural development. In some areas, this has changed the character of the wetland from seasonally flooded to semipermanent water and changed all the organisms and birds therein. Other areas were exposed for sugar cane and other crops and have demonstrated major subsidence because of the decomposition of organic matter. Enrichment and sedimentation have been major problems on large river impoundments, many of which function as wetlands in the more shallow portions. In addition, poor land management upslope has exacerbated the problem, seriously affecting all aquatic resources and waterbirds in riverine wetlands (Bellrose, Paveglio and Steffeck 1979) and estuaries (Perry and Deller 1996).

Although river impoundments currently are less commonly supported by government agencies in developed nations, the economies of some nations seem focused on construction of ever-more massive impoundments (China, Canada), reflected also in long-range planning (South America and Africa). This may bring more water, electrical power, and perhaps food for people, agriculture, and industry, but it will flood wetlands and other lowland natural habitats and will stabilize water levels with a strong probability of a net reduction in productivity of most water resources.

Coastal tidal impoundments dedicated to expanding reclamation of land for agriculture or for energy projects are increasing as land values rise and technology allows; these will have major impacts on saltmarsh and mudflats vital to many local and global bird species (Goss-Custard and Yates 1992).

14.2 Human impacts on existing wetlands

In addition to direct losses, many indirect and modifying influences have influenced bird-breeding areas, migration stop-over sites, or wintering areas. Wetlands regularly are affected by water diversion, which may result in drying of interior freshwater areas or modification of salinity balance in coastal wetlands or interior sinks, and by upslope practices such as overgrazing or tilling that influence sedimentation and flooding (e.g., the Mississippi River basin). Changes in water regimes caused by irrigation withdrawal and flooding has a drastic impact on birds, and typically would eliminate nesting, as shown from observations on drought and natural flooding cycles (Weller 1988, Weller, Wingfield and Low 1958). Because of their mobility, some birds do still use these areas for resting and feeding but must seek nest sites elsewhere. These influences, plus direct filling, also eliminate smaller basins and reduce diversity within wetland complexes (e.g., Prairie Potholes of North America)(Weller 1988).

Multiple effects of pollutants on waterfowl food plants have seriously affected duck populations in the Chesapeake Bay (Perry and Deller 1996). Ultimately, these community changes influence resource availability for waterbirds that feed upon them. Many industrial chemicals and fertilizers cause problems when drainage occurs from urban settings; high concentrations of elements such as selenium can occur in irrigated farmland of arid regions (National Research Council 1989); and acidification can result from mining and industrial products. These are all potential stresses on wetland systems.

Fire and grazing may be no less drastic but perhaps are less permanent in damage because they are natural processes common to basin wetlands and hence adapted to by many plants. Nevertheless, management strategies based on bird use of the areas are essential, including timing of fires and grazing and setting a suitable grazing intensity to avoid serious physical impact on the vegetation (Gilbert *et al.* 1996, Weller 1996). Intense grazing tends to eliminate vegetation and induce sedimentation, favoring species that like mudflats, meadows, and open water, thereby changing the bird assemblage. However, severity of sedimentation often is relatively less severe under controlled grazing than upslope tilling (Luo *et al.* 1997).

There have been some major and many seemingly minor but ubiquitous dangers that confront waterbirds daily. Highly toxic and long-lived pesticides were early recognized as serious in waterbirds because of their concentration especially in fish of coastal marine systems. As a result, of this concentration Brown Pelicans in North America suffered major losses and were nearly eliminated from the Gulf of Mexico and the Pacific Coast breeding areas, whereas interior- nesting White Pelicans in non-agricultural areas were less affected. Although many of the more serious pesticides have been banned or closely regulated in many countries, others are widespread in third-world nations. Interior basins fed by large streams, such as the world's largest freshwater wetland, the Pantanal of Brazil (Alho, Lacher and Gonclaves 1988), have suffered from heavy use of agricultural pesticides, which has resulted in modified fish and invertebrate communities.

Oil and gas development continues, especially in remote and often delicate ecosystems around the world, and there are numerous cases of small but locally serious impacts on birds (see Burger 1997). Waste deposition of oil, caustic salts, chemicals, and other substances harmful to birds are surprisingly widespread and poorly controlled and very attractive to waterbirds when in a fluid state.

Line fishing with exposed hooks is deadly to diving birds in fresh and salt-water, and net fishing can be serious for diving birds like loons, grebes, and ducks (Bartonek 1965, Schorger 1947). Because of nocturnal migration, often over land, collisions with human structures is common, including fences and telephone lines (Cornwell 1971) and powerlines (Malcolm 1982). Although lead shotgun pellets have been banned for hunting waterfowl over water in the USA, the residual lead in the substrates still accounts for 3% of waterfowl mortality in the whole country, and more in localized areas (Anderson and Havera 1989, Sanderson and Bellrose 1986). Any species that strains bottom oozes for grit is subject to lead intake, as recorded in flamingos in Mexico and Europe (Schmitz et al. 1990). Moreover, the use of lead for shotguns is still widespread in many areas of the world.

Botulism and other natural diseases are still poorly understood but strike unexpectedly, with huge and uncontrollable mortality, as in the loss of 10 000 pelicans in the Salton Sea of California (Anonymous 1997), an area where in past years major losses of grebes have occurred. Bird–aircraft strikes are still more common than they should be and often involve waterbirds because of their flock and roosting behavior (Holden 1997), or because they become concentrated in certain areas in response to societal strategies like landfills (gulls) or protected breeding areas (Canada Geese, Cooper 1997).

Fish production, sport fishing, and introduction of exotic fish have resulted in direct impacts on habitat through introduction of herbivores like carp, and through competition for food of native birds (Bouffard and Hanson 1997, Weller 1994) from exotics and trout reared in wetlands. A related topic is the intentional persecution of cormorants and other fish-eating birds worldwide because these birds are viewed as competitive with fishermen (Nettleship and Duffy 1995). Few studies of free-living birds have shown a serious impact on fish stocks, but impacts on concentrations of catfish or other pond-production species obviously have been greater. Realistically, we expect too much when we think that free-living birds long-adapted to mobile resource exploitation will ignore this specially prepared feast!

The potential relationship between finfish and shellfish harvest and waterbird populations that feed on these resources seems obvious but is unmeasured. The recent crash in the population of Horseshoe Crabs at a major spring stopover in New Jersey is of great concern because of the importance of their spawn as food for migrant shorebirds; populations of the Red Knot at some key stopovers were at a lower level in 1997 than in the prior 10 years, and commercial harvest of crabs has been banned in an effort to enhance crab populations (Drennan 1997).

Several issues of concern are more difficult to assess, and only monitoring over time will demonstrate their seriousness. Acid precipitation seems to be affecting bird rearing success in northern areas, mainly through reduction of food resources such as invertebrates (Blancher and McNicol 1991, Des Granges and Hunter 1987). If directional global climate change is real, mathematical models predict significant effects on vegetation, water regimes, and, presumably, waterbirds (Poiani and Johnson 1991, Poiani *et al.* 1996). However, these are expected to differ by region and by hemisphere.

14.3 Direct disturbance of waterbirds in remaining wetlands

Studies of disturbance of waterbirds by human intrusions have shown a variety of responses to and impacts of such intrusion (Burger 1981, Dalhgren and Korschgen 1992, Klein 1993, Rogers and Burger 1981, Titus and van Druff 1981). In situations where people visit commonly or continuously (e.g., birdwatching from a distance) without direct harm (e.g., hunting) to birds or with direct disturbance (walking, running, boating), birds seem to habituate to some types of disturbance. One suspects that residents or returning young of prior years probably constitute the braver individuals, and that the resources at the site are worth the risk. In areas where migrants constitute the majority, birds seem less brave and many birds seem to choose less-humanized areas by avoidance (Klein, Humphrey and Percival 1995). The more significant impacts seem to result from continuous intrusion from motorized vehicles like boats and planes, which prevent normal feeding and resting activities. Species like Black Brant, which fear airplanes (especially helicopters), do not seem to adjust, and habitat use can be severely reduced by such activity. Behavioral studies demonstrate that birds do avoid people and do expend extra energy to do so, and this effort could affect body condition, as predicted by computer modeling of the severity and incidence of disturbance (Miller 1994).

Observations of recreational activities such as use of motorized toys near waterbirds have shown mixed responses, with model boats less disturbing (Bamford, Davies and van Delft 1990) than model airplanes (Harrington 1996). A strategy to limit the areas of disturbance in or near refuges and preserves is becoming recognized as an essential part of a management strategy where waterbirds and society coexist. This may take the form of closures of refuge areas to boating or tourists during nesting or other sensitive periods, and the use of buffer zones whenever needed (Rogers and Smith 1997).

These are but a few of the problems facing wetland birds directly or their habitats. Conservationists are not without challenges!

14.4 Unwanted biodiversity

Society has had another major impact on avian diversity and plant composition in wetlands, as in other habitats, by directly importing foreign (exotic) birds and other animals and plants, or by inducing unwanted mobility or range expansion via drainage and other land-use changes. As a result, some wetland birds occur now where they once did not. Most conservationists strongly oppose intentional releases of exotic birds because of the potential for interspecific competition with impact on native species, such as the classic examples of the Starling, House Sparrow, and Common Pigeon. But such events have occurred with wetland species, sometimes with the approval of well-meaning but poorly informed conservationists. Most intentional releases have been of waterfowl intended for either hunting or esthetic purposes. For example, introduction of Mallard ducks occurred for hunting purposes in many areas: in New Zealand they hybridized with the native Grey Duck, in Hawaii they interbred with Hawaiian Ducks (Weller 1980), and in the eastern USA interbreeding with local Black Ducks is widespread. Fortunately, Mallards introduced in south Texas, where potential interbreeding with local Mottled Ducks might have been a problem, have not survived well and now are rare (Kiel 1970). They also have disappeared from the Falkland Islands, where competition with other dabbling ducks might have been a problem, but they have been introduced into Argentina.

For esthetic purposes, Mute Swans from Europe have been introduced or escaped from captivity widely around the world because of their use in cemetery lakes for atmosphere. Free-living populations now occur in Toronto, the northwestern and eastern coasts of the USA, and in New Zealand, to mention but a few. At least one study has found no apparent negative impacts on local species (Conover and Kania 1994). Black Swans have been introduced from Australia to New Zealand and have been very successful and are now abundant there. A similar situation holds for Canada Geese intentionally introduced and now widespread in New Zealand, but opinion varies between hunters, farmers, and conservationists as to their merits. In North America, Canada Geese released in urban areas and refuges for restoration and esthetic purposes have been protected by lack of hunting and artificial feeding so that they become nuisance birds. Moreover, such introductions have often resulted in mixing of genetic strains, which has confounded the original distribution patterns of geographic races.

The desire of aviculturists to keep and rear exotic waterfowl has had mixed blessings. Birders now constantly report accidental or very rare ducks more commonly, and escapes from captivity are suspected in many cases. The magnificent Mandarin Duck of the Orient has been a favorite show bird in many collections and escaped from captivity in England long ago. It now is free-living over a wide area there and more recently has been free-living in California as well (Shurtleff and Savage 1996). Fortunately, it does not seem to hybridize with its closest relative, the North American Wood Duck. A more serious recent example of the potential consequence of escapees is the North American Ruddy Duck in England. Its beautiful colors and unique behavior made it an attractive bird to show; the challenge of rearing young resulted in captive birds being imported to the UK, where they subsequently escaped in 1952. By 1992, the Ruddy Duck had spread to 15 European countries and populations are still steadily increasing. Unfortunately, it is an aggressive species and cross-mates with the European White-headed Duck (another stifftail), which is already in trouble because of habitat loss (Johnsgard and Carbonell 1996).

As mentioned above, the natural range of the Cattle Egret is now nearly worldwide in tropical and warm-temperate areas. Apparently, efforts have been made to introduce it on islands like Hawaii and elsewhere, but most of its success seems to result from its own pioneering behavior. As its name implies, it feeds at the feet of feeding livestock and large wild ungulates, where it gathers insects and other small prey stirred by their movement. It also feeds in newly plowed fields, often following the tractor like gulls and harriers. Therefore, it occupies a new or expanded habitat created by agriculture. It has since expanded into heronries throughout the USA and makes an annual migration into Mexico, a pattern than has evolved since the early 1950s. Although it does not seem to compete directly for food (at least with wetland species), some observers speculate that its dominance of nest sites could adversely affect other species in mixed colonies.

It is obvious that introductions can produce serious direct results and have the potential of carrying diseases and competing in ways we cannot easily measure. Moreover, it modifies natural and long-established bird communities in ways that we have not yet adequately assessed. Although such releases may be interesting experimentally, they are poorly controlled and should not be condoned. The International Treaty on Biodiversity incorporates a statement on limits and controls of invaders and is encouraging prompt and meaningful policies and control methods (Baskin 1996).

Indirectly, introductions of other exotic species can have important influences on native waterbirds. The accidental introduction of the Zebra Mussel into Europe and the USA is having major impacts on the trophic ecology of many fish and aquatic birds. Diving ducks use them readily and populations of some species may benefit (Custer and Custer 1996, Hamilton and Ankney 1994), but far more serious implications have been postulated for these aquatic communities (Ludyanskiy, McDonald and MacNeil 1993). Introduced fish have had effects on many natural systems around the world, and Nutria from South America and Muskrats from North America have had an impact on natural vegetation and crops in many areas and also displaced native species.

There are currently major concerns about wetland plant species that have been moved around the world and are having serious impacts on natural plant communities and, thereby, wetland birds. Some may show positive effects, as in waterfowl use of the introduced *Hydrilla* spp. (Esler 1990) but many have had major negative impacts on plant dominance and plant management problems that influence birds and bird use of habitats (Perry and Deller 1996). This problem has been recognized for a long time but is getting worse; it is finally attracting enough concern that it can be dealt with as a matter of international policy (Vitousek *et al.* 1996).

References

Alho, C. J. R., Lacher, T. E. Jr, and Gonclaves, H. C. (1988). Environmental degradation in the Pantanal Ecosystem. *BioScience* 38, 164–71.

Anderson, W. L. and Havera, S. P. (1989). Lead poisoning in Illinois waterfowl (1977–1988) and the implementation of nontoxic shot regulations. *Illinois Natural History Survey Biological Notes* 133, 1–37.

Anonymous (1997). Pelican Peril. *Birder's World* 11, 12.

Bamford, A. R., Davies, S. J., and van Delft, R. (1990). The effects of model power boats on waterbirds at Herdsman Lake, Perth, Western Australia. *Emu* 90, 260–5.

Bartonek, J. C. (1965). Mortality of diving ducks on Lake Winnipegosis through commercial fishing. *Canadian Field-Naturalist* 79, 15–20.

Baskin, Y. (1996). Curbing undesirable invaders. *BioScience* 46, 732–6.

Bellrose, F. C., Paveglio, F. L. Jr, and Steffeck, D. W. (1979). Waterfowl populations and the changing environment of the Illinois River Valley. *Illinois Natural History Survey Bulletin* 32, 1–54.

Blancher, P. J. and McNicol, D. K. (1991). Tree swallow diet in relation to wetland acidity. *Canadian Journal of Zoology* 69, 2629–37.

Bouffard, S. H. and Hanson, M. A. (1997). Fish in waterfowl marshes: waterfowl managers' perspective. *Wildlife Society Bulletin* 25, 146–57.

Burger, J. (1981). The effect of human activity on birds at a coastal bay. *Biological Conservation* 21, 231–41.

Burger, J. (1997). Effects of oiling on feeding behavior of sanderlings and semipalmated plovers in New Jersey. *Condor* 99, 290–8.

Conover, M. R. and Kania, G. S. (1994). Impact of interspecific aggression and herbivory by Mute Swans on native waterfowl and aquatic vegetation in New England. *Auk* 111, 744–8.

Cooper, J. A. and Keefe, T. Managing urban Canada Geese: policies and procedures. *Transactions of the North American Wildlife & Natural Resources Conference* 62, 412–30.

Cornwell, G. (1971). Collisions with wires – a source of anatid mortality. *Wilson Bulletin* 83, 305–6.

Custer, C. M. and Custer, T. W. (1996). Food habits of diving ducks in the Great Lakes after the zebra mussel invasion. *Journal of Field Ornithology* 67, 86–99.

Dahl, T. E. (1990). *Wetland losses in the United States 1780s to 1980s.* Washington DC: US Fish & Wildlife Service.

Dahl, T. E., Johnson, C. E., and Frayer, W. E. (1991). *Wetlands status and trends in the coterminous United States mid-1970s to mid 1980s.* Washington DC: US Fish and Wildlife Service.

Dahlgren, R. B. and Korschgen, C. E. (1992). *Human disturbance of waterfowl: an annotated bibliography.* Resource Publication 188. Washington DC: U S Fish & Wildlife Service.

Des Granges, J. L. and Hunter, M. L. Jr. (1987). Duckling response to lake acidification. *Transactions of the North American Wildlife & Natural Resources Conference* 52, 636–44.

Douglas, M. S. (1947). *The everglades: river of grass.* St Simons, GA: Mockingbird Books.

Drennan, S. R. (1997). Horseshoe crab decline in Delaware Bay. *National Audubon Society Field-Notes* 51, 6.

Dugan, P. J. (ed.) (1990). *Wetland conservation, a review of current issues and required action.* Gland, Switzerland: World Conservation Union, IUCN.

Earnest, A. (165). *The art of the decoy: American bird carvings.* New York: Bramhall House.

Esler, D. (1990). Waterfowl habitat use on a Texas reservoir with Hydrilla. *Proceedings of the Annual Conference of the Southeastern Association of Fish and Wildlife Agencies* 1990: 390– 400.

Forbush, E. H. (1916). *A history of the game birds, wild-fowl and shore birds of Massachusetts and adjacent states,* 2nd edn. Boston: Massachusetts State Board of Agriculture.

Gilbert, D. W., Anderson, D. R., Ringelman, J. K., & Szymczak, M. R. (1996). Response of nesting ducks to habitat management on the Monte Vista National Wildlife Refuge, Colorado. *Wildlife Monographs* 131, 1–44.

Good, R. E., Whigham, D. F., and Simpson, R. L.(eds.) (1979). *Freshwater wetlands, ecological processes and management potential.* New York: Academic Press.

Goss-Custard, J. D. and Yates, M. G. (1992). Towards predicting the effect of salt-marsh reclamation on feeding bird numbers on the Wash. *Journal of Applied Ecology* 29, 330–40.

Gosselink, J. G., Odum, E. P,. and Pope, R. M. (1973). *The value of a tidal marsh.* Baton Rouge, LA: Center for Wetland resources, Louisiana State University.

Greeson, H., Clark, J., and Clark, J. (eds.) (1978). *Wetland functions and values: the state of our understanding.* Minneapolis, MN: American Water Resources Association.

Hamilton, D. J. and Ankney, C. D. (1994). Consumption of zebra mussels *Dreissena polymorpha* by diving ducks in Lakes Erie and St Clair. *Wildfowl* 45, 159–66.

Harrington, B. (1996). *The flight of the Red Knot.* New York: W. W. Norton,

Holden, C. (1997). Eco-solution to airport bird pests. *Science* 275, 487.

Johnsgard, P. A. and Carbonell, M. (1996). *Ruddy ducks and other stifftails.* Norman, OK: University of Oklahoma Press.

Kiel, W. H. Jr (1970). *A release of hand-reared mallards in South Texas.* Publication MP968. College Station, TX: Texas A&M University Agricultural Experiment Station.

Klein, M. L. (1993). Waterbird behavioral response to human disturbances. *Wildlife Society Bulletin* 21, 31–9.

Klein, M. L., Humphrey, S. R., and Percival, F. (1995). Effects of ecotourism on distribution of waterbirds in a wildlife refuge. *Conservation Biology* 9, 1454–65.

Krapu, G. L. (1996). Effects of a legal drain clean-out on wetlands and waterbirds: A recent case history. *Wetlands* 16, 150–62.

Larson, C. S. and Kelly, R. L. (1995). Bioarchaeology of the Stillwater Marsh, prehistoric human adaptations in the western Great Basin. *American Museum of Natural History, Anthropology Papers* 77, 1–170.

Ludyanskiy, M. L., McDonald, D., and MacNeil, D. (1993). Impact of the zebra mussel, a bivalve invader. *BioScience* 43, 533–44.

Luo, H., Smith, L. M., Allen, B. L., and Haukos, D. A. (1997). Effects of sedimentation on playa wetland volume. *Ecological Applications* 7, 247–52.

Malcolm, J. M. (1982). Bird collisions with a power transmission line and their relation to botulism at a Montana wetland. *Wildlife Society Bulletin* 10, 297–304.

Maltby, E. (1986). *Waterlogged wealth.* London: Earthscan. International Institute for Environment and Development.

Maxwell, G. (1957). *People of the reeds.* New York: Harper and Row.

Miller, M. W. (1994). Route selection to minimize helicopter disturbance of molting Pacific Black Brant: a simulation. *Arctic* 47, 341–9.

Miller, M. W. and Nudds, T. D. (1996). Prairie landscape change and flooding in the Mississippi River Valley. *Conservation Biology* 10, 847–53.

Myers, J. P., Morrison, R. G., Antas, P. Z., Harringon, B. A., Lovejoy, T. E., Sallaberry, M., Senner, S. E., and Tarak, A. (1987). Conservation strategies for migratory species. *American Scientist* 75, 18–26.

National Research Council (1989). *Irrigation-induced water quality problems.* Washington, DC: National Research Council.

Nettleship, D. N. and Duffy, D. C. (eds.) (1995). The Double-crested Cormorant: biology, conservation and management. *Colonial Waterbirds* 18, 1–256.

Perry, M. C. and Deller, A. S. (1996). Review of factors affecting the distribution and abundance of waterfowl in shallow-water habitats of Chesapeake Bay. *Estuaries* 19, 272–8.

Poiani, K. A. and Johnson, W. C. (1991). Global warming and prairie wetlands. *BioScience* 41, 611–18.

Poiani, K. A., Johnson, W.C., Swanson, G. A., and Winter, T. C. (1996). Climate change and northern prairie wetlands: simulations of long-term dynamics. *Limnology & Oceanography* 4l, 871–81.

Rogers, J. A. Jr, and Burger, J. (eds.) (1981). Symposium on human disturbance and colonial waterbirds. *Colonial Waterbirds* 4, 1–70.

Rogers, J. A. and Smith, H. T. (1997). Buffer zone distances to protect foraging and loafing waterbirds from human disturbance in Florida. *Wildlife Society Bulletin* **25**, 139–45.

Sanderson, G. C. and Bellrose, F. C. (1986). A review of the problem of lead poisoning in waterfowl. Urbana, IL: *Illinois Natural History Survey Special Publication* **4**, 1–34.

Schmitz, R. A., Aguirre, A. A., Cook, R. S., and Baldassarre, G. (1990). Lead poisoning of Carribean flamingos in Yucatan, Mexico. *Wildlife Society Bulletin* **18**, 399–404.

Schorger, A. W. (1947). The deep diving of the Loon and the Old-squaw and its mechanism. *Wilson Bulletin* **59**, 151–9.

Shurtleff, L. L. and Savage, C. (1996). *The wood duck and the mandarin.* Berkeley, CA: University of California Press.

Tiner, R. W. Jr (1984). *Wetlands of the United States: current status and recent trends.* Washington DC: U S Fish & Wildlife Service.

Titus, J. R. and Van Druff, L. W. (1981). Response of the common loon to recreational pressure in the Boundary Waters Canoe Area, northeastern Minnesota. *Wildlife Monographs* **79**, 1–58.

Vitousek, P. M., D'Antonio, C. M., Loope, L. L., and Westbrooks, R. (1996). Biological invasions as global environmental change. *American Scientist* **84**, 468–78.

Weller, M. W. (1980). *The island waterfowl.* Ames, IA: Iowa State University Press.

Weller, M W. (1988). Issues and approaches in assessing cumulative impacts on wildlife habitat in wetlands. *Environmental Management* **12**, 695–701.

Weller, M. W. (1994). *Freshwater marshes*, 3rd edn. Minneapolis, MN: University of Minnesota Press,

Weller, M. W. (1996). Birds of rangeland wetlands. In *Rangeland Wildlife*, ed. P. Krausmann, pp. 71–82. Denver, CO: Society for Range Management.

Weller, M. W., Wingfield, B. H., and Low, J. B. (1958). Effects of habitat deterioration on bird populations of a small Utah marsh. *Condor* **60**, 220–6.

Further reading

Bratton, S. P. (1990). Boat disturbance of Ciconiiforms in Georgia estuaries. *Colonial Waterbirds* **13**, 124–5.

McNikol, D. K., Bendell, B. E., and Ross, R. K. (1987). Studies of the effects of acidification on aquatic wildlife in Canada: waterfowl and trophic relationships in small lakes in northern Ontario. Ottawa: *Canadian Wildlife Service Occasional Papers* **62**, 1–76

Parker, G. R., Petrie, M. J. and Sears, D. T. (1992). Waterfowl distribution relative to wetland acidity. *Journal of Wildlife Management* **56**, 268–74.

Shoemaker, T. G. (1989). *Wildlife and water projects on the Platte River.* Wildlife Report 1988/89. New York: National Audubon Society.

Vos, D. K., Ryder, R. A. and Graul, W. D. (1985). Response of Great Blue Herons to human disturbance in Northcentral Colorado. *Colonial Waterbirds* **8**, 13–22.

Weller, M. W., Jensen, K. C., Taylor, E. J., Miller, M. W., Bollinger, K. S., Derksen, D. V., and Esler, D. (1994). Assessment of shoreline vegetation in relation to use by molting black brant *Branta bernicla nigricans* on the Alaskan coastal plain. *Biological Conservation* **70**, 219–25.

15

Conservation and management strategies

The existing problems that wetland birds face and the potential applications of ecological data must be considered in the light of current structure and approaches in wildlife conservation. Historically, conservation and management efforts have focused on **legal protection, habitat management, population management,** and **human dimensions,** and all must be used concurrently to be effective. Our focus here is habitat management, which has been used for different purposes but is often aimed at **species management** when, at least for wetlands, most management influences the entire community.

Early efforts of habitat preservation in the USA, and probably in most countries, focused on wetland habitat protection. Direct purchase for refuges, sanctuaries, and preserves has been common, but leasing and legally binding conservation easements and agreements have been more common recently as a means of controlling more areas at lower cost per unit area, especially when multiple ownership is involved. Protection of small basins or patches of habitat has worked well for wetlands because many birds are habitat-island species that use several habitat units and fly between them. In the USA, there are over 500 National Wildlife Refuges operated by the US Fish and Wildlife Service involving over 92 million acres, most of which include major wetlands (Riley and Riley 1992). Other federal agencies such as the Forest Service and Bureau of Land Management also hold title to and manage similar areas, and the Natural Resource Conservation Service program accomplishes similar protection on private lands through the Soil Conservation Reserve Program (Johnson and Igl 1995). Many states, counties and cities have protected additional and generally smaller but no-less valuable areas through refuges, public hunting areas, nature centers, and green belts. Similar strategies have occurred around the world, but, in most areas, these constitute a small percentage of the original area that once supported wetland bird diversity and numbers. However, these wetlands often lack surrounding upland vegetation so important to water quality (e.g., sediment control) and for nesting habitat for birds that use wetland margins. Moreover, they tend to represent the typical or expected examples rather than the diversity of habitats essential to attracting different species of bird.

Land acquisition for conservation of wildlife sometimes has involved landowners, sometimes has resulted from finances and support of non-owners, and sometimes agencies have merely moved in and bought all that was possible. In some areas, these approaches have worked, and sometimes they have produced resistance and controversy and have been more harmful than good (e.g., affecting local tax bases). Therefore, greater emphasis is now on "team-building" and "partnerships" to insure cooperation of land owners and land managers to develop shared responsibility for resources, greater sustainability of resources used by humans and wildlife and economic benefits for land owners. With wetlands in particular, we have learned slowly that watersheds that influence water quality may be as important as the water-collecting basin, and yet ownership of all is impractical. Therefore, cooperative, multipurpose, sustainable watershed management programs are essential (Dugan 1990). Some agencies and organizations pay land taxes to alleviate problems of excess ownership in certain high-value areas (e.g., waterfowl refuges and production areas). Managers and guardians work hand-in-hand with civic groups to promote recognition of the importance of wildlife in the local economy, for example British Columbia's Black Brant Festival, Oregon's Shorebird Festival, and New Mexico's Crane Festival (National Fish and Wildlife Foundation 1997); similar waterbird events in the UK are sponsored by the Waterfowl and Wetlands Trust.

15.1 Habitat management approaches

Management has become an unpopular word to many who prefer naturally maintained and sustained areas, and many have returned to the use of the word conservation. I use the term management to imply an active conservation strategy that may be aimed at habitats or fauna but in which policy decisions *not* to do something is part of the approach. After years of acquisition of natural areas, the Nature Conservancy came to recognize that wild areas often need management by fire, fencing, or even more aggressive approaches because the "island" to be conserved has and will continue to be modified by external forces. Wetlands in particular are modified by upslope use (and especially abuse), and these areas often become the focus of wetland restoration efforts. Moreover, short-term succession is such a driving force that we may lose the habitat features and values we wish to preserve unless we (i) maintain those driving and dynamic natural forces characteristic of the landscape especially water regimes, and (ii) compensate for abnormal patterns produced by impinging human activities. Moreover, wetland management often involves maintenance of natural variation and near-catastrophic extremes rather than striving for regularity as occurs in some systems.

There are many publications on habitat management for wetland wildlife, cited above, which describe approaches to manipulating the ecosystem for the bird community. My purpose is not to repeat those here but to comment on selected issues as examples of policy and current actions. Most current wetland-management strategies are based upon manipulation of processes that drive the system, such as water timing and dynamics that result in changes in the direction rates of plant and animal succession (Fredrickson and Reid 1990, Fredrickson and Taylor 1982, Payne 1992, Scott 1982, Weller 1978, 1990). Most support the general philosophy on use of ecologically sound approaches to conserve and manage habitat for wetland wildlife.

Although the concept of the ecosystem has been around for a long time, the move of applied natural resource scientists from species-oriented to community management to **ecosystem management** has been gradual and controversial (e.g., Stanley 1996). The current literature often uses the term ecosystem management, but usually we are dealing with a "community" because we often do little to understand, measure, and manipulate the physical components and processes. Nevertheless, use of the term ecosystem forces us into system-level thinking, and the recognition that manipulations influence all components and processes within the system.

Few biological systems so dramatically respond to a single driver, water, as do wetlands. All organisms within the ecosystem respond, making a holistic perspective and management strategy essential. In the past, we have tended to manage for a specific bird because of interests such as hunting or esthetic preference, but, whatever the goals, processes that result in sustainable, self-maintaining wetlands should be a primary strategy. Intensive management that requires repetition is costly, and results often are unpredictable.

The development of **island biogeography theory** (MacArthur and Wilson 1967) led to designs for natural preserves (Diamond 1975) that dealt with the conceptual framework and the selection of "habitat islands" of habitats of varying sizes, distances, and distribution for various plants and animals. Many wetlands are islands of habitat and, therefore, can be considered in this perspective (Weller 1979); the acquired refuges and preserves have contributed in a major but unmeasured way in protecting wetland birds and ensuring diversity because they fit within the strategy that has evolved as a way-of-life for many mobile birds. Assurance of water and food has been a product of a mobile search pattern, and these seemingly isolated areas can attract an amazing diversity and number of birds that varies temporally. To be sure, some areas do not seem to be discovered quickly by birds nor used to the degree we think possible, and this may be related to migratory corridors, or to small size and isolation. Moreover, it is important to recognize that wetlands are not as truly isolated as are oceanic islands, and many external forces such as water sources and land-use influence their functioning (Kushlan 1979).

Other wetland species require more **linear habitats,** which form pathways or corridors, as noted in coastal or riverine birds. While it generally is not possible to protect continuous habitat, we have at least recognized that birds often fly awhile and feed awhile as they move more slowly along these contiguous habitats. Preserving many small areas distributed in a linear pattern may be more important than putting all our eggs in one or a few baskets. However, it is essential to preserve major migration stops for large numbers of long-range migrants as well, and the international Shorebird Reserve Program has identified and is working toward such measures. Other examples exist throughout the world, reflecting the interest and zeal of amateur as well as professional bird students.

The term **biodiversity** has been applied broadly to an overall conservation goal for society and as a technique for identifying key areas based on species richness of all taxonomic entities and the supporting ecosystem structure (Noss and Cooperrider 1994). Computer-based geographic information systems and data bases on species distribution allow the identification of species-rich areas that have few or no preserves, an approach termed **gap analysis** (Kiester *et al.* 1996, Scott *et al.* 1993). This system is also being used on wetland animals, but it seems better suited to larger scales (Noss and Cooperrider 1994). A currently popular index for selecting important areas to conserve is based on species-rich areas (often termed "hotspots") (Dobson *et al.* 1997), but such areas are more difficult to identify when dealing with variable environments like wetlands and migratory or even nomadic species. Not everyone is convinced that simple measures of biodiversity are adequate for major conservation decisions that could affect worldwide conservation policies. Part of the weakness of the system seems to be that ecosystems are not consistent in the relationship of number of species to uniqueness, and with wetlands, they are not constant over time. Several studies have concluded that assessing the priority of differing wetlands for preservation based on species richness may not identify the important habitats for those bird species that have narrow or precise requirements, and for which no other habitats are suitable (Burger, Schisler and Lesser 1982, Weller 1994). Moreover, because many wetlands vary markedly in water regimes, they may be under-valued if assessed by a few surveys at a poor time when they actually fill an important need such as for roosting habitat, with other functions such as feeding or breeding occurring elsewhere. Hence, our analyses often must treat wetlands as separate habitat units that serve a bird species or group, recognizing the spatial and temporal aspects of these components of an annual habitat that fulfill different functions and in different areas. Additionally, some workers stress that preserving a natural area of any size is still an ideal goal when no other such natural area exists (Shafer 1995) – as is often the case for isolated or urban wetlands.

Once an area of habitat is acquired, management too-often fails because the processes that drive the system often are unknown, or management strategies are applied in a rigid manner. In my experience, wetlands with water-control structures tend to have water gauges, whereas naturally controlled ones rarely do even if under study. Without being able to relate wetland vegetation and bird use to water regimes, we cannot understand regulators in the system, we have no measure of suitable and optimal conditions, and, therefore, we cannot set realistic goals or measure our success at meeting these. Another failing has been what I call the "pick-up truck syndrome," which induces "a view from afar" when one needs a "hip- boot approach." Regular monitoring and corrective action are essential. Moreover, with wetlands, this requires a longer-term view than in terrestrial systems, because water regulation must be planned well in advance to induce successional changes that lead to changes in bird habitat (Weller 1978). Such **adaptive management** must be dynamic, changing as needs demand to meet stated goals. However, it involves carefully planned objectives, experimental approaches to ensure reasonable answers, and a monitoring system than provides feedback and allows change toward a better final product (Holling 1978, Walters 1986).

15.1.1 AVIAN HABITAT MANAGEMENT IN EXISTING WETLANDS

Wetland management for birds and other wildlife must take a community and ecosystem perspective, recognizing the influences of most actions on species that are not major targets. Moreover, process-oriented management involving drivers of the system (e.g., water regime and fire) that influence nutrient dynamics and natural plant succession is the most ecologically and economically sound approach. In addition to the manipulation of physical processes, grazers like Muskrats, Nutria, geese, Beaver, Marsh Deer, Water Buffalo, and domestic livestock can have major impacts (Weller 1996). Large numbers of herbivorous fish alter vegetation, and predatory fish modify invertebrate populations and thereby bird use, but this is less measurable than are the positive effects of large numbers of fish available as food for egrets and herons. Undoubtedly, these food-web interactions are essential when devising techniques needed to manipulate wetlands for optimal diversity and production.

15.1.2 RESTORATION, ENHANCEMENT, AND CREATION OF WETLANDS

Much of the management of wetlands today, regardless of objective and wetland type, is an effort to improve habitat within and around the wetland for maximal production and diversity of birds or other wildlife. This is termed **enhancement** (Weller 1990) and may involve manipulating water regimes, water quality, and vegetation through methods described above. Such actions may be legally demanded in cases of mitigation for wetland losses, but in most cases, the approach is used by private, state, and federal managers attempting to compensate for lost or impacted wetlands nearby. The techniques may be less drastic than in restoration, but the degree of improvement also is less dramatic and, therefore, not easily measured.

The most dramatic and measurable results are seen when returning a former wetland to its previous productive state, termed **restoration**. In some cases, the heritage of the area may only be discernable via the presence of a wetland **seed bank**, a pool of long-lived residual seeds in the substrate that will germinate when water is returned and maintained at a suitable level (van der

■ CHANNELIZED

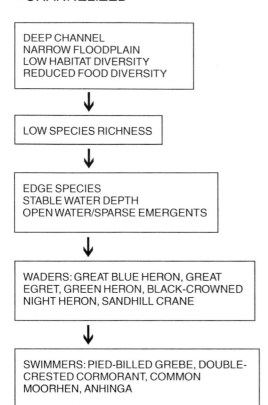

DEEP CHANNEL
NARROW FLOODPLAIN
LOW HABITAT DIVERSITY
REDUCED FOOD DIVERSITY

↓

LOW SPECIES RICHNESS

↓

EDGE SPECIES
STABLE WATER DEPTH
OPEN WATER/SPARSE EMERGENTS

↓

WADERS: GREAT BLUE HERON, GREAT
EGRET, GREEN HERON, BLACK-CROWNED
NIGHT HERON, SANDHILL CRANE

↓

SWIMMERS: PIED-BILLED GREBE, DOUBLE-
CRESTED CORMORANT, COMMON
MOORHEN, ANHINGA

■ RESTORED/NATURAL

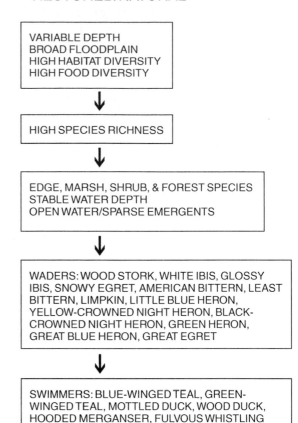

VARIABLE DEPTH
BROAD FLOODPLAIN
HIGH HABITAT DIVERSITY
HIGH FOOD DIVERSITY

↓

HIGH SPECIES RICHNESS

↓

EDGE, MARSH, SHRUB, & FOREST SPECIES
STABLE WATER DEPTH
OPEN WATER/SPARSE EMERGENTS

↓

WADERS: WOOD STORK, WHITE IBIS, GLOSSY
IBIS, SNOWY EGRET, AMERICAN BITTERN, LEAST
BITTERN, LIMPKIN, LITTLE BLUE HERON,
YELLOW-CROWNED NIGHT HERON, BLACK-
CROWNED NIGHT HERON, GREEN HERON,
GREAT BLUE HERON, GREAT EGRET

↓

SWIMMERS: BLUE-WINGED TEAL, GREEN-
WINGED TEAL, MOTTLED DUCK, WOOD DUCK,
HOODED MERGANSER, FULVOUS WHISTLING
DUCK, RING-NECKED DUCK, COOT, COMMON
MOORHEN

Valk and Davis 1978). The process is relatively simple in basin wetlands, through return of normal hydrologic regimes to the area. Results are not totally predictable, but the effort often is highly successful at re-establishing vegetation. Return of other organisms that influence bird use is more variable. When foods like terrestrial seeds or invertebrates are present and flooded, bird response can be immediate. The response of breeding birds occurs more slowly unless there are nearby wetlands to attract them, but occasionally birds breed within several years. In river systems, restoration involves either creating or reopening old channels, oxbows, and backwater areas to trap and hold water, which then attracts more diverse species that use such habitats as opposed to channels (Fig. 15.1).

Creation of wetlands is less common than the above approaches because it is more costly, more demanding of intensive management, and less likely to succeed than restoration. But it is becoming more common because develop-

Figure 15.1. Conceptual model of predicted responses of restoration actions on bird species composition of the now-channelized Kissimmee River (Weller (1995) Use of two waterbird guilds as evaluation tools for the Kissimmee River restoration. *Restoration Ecology* 3, 211–14. Reprinted by permission of Blackwell Science, Inc.).

ers often are willing to pay the high price to mitigate for wetlands destroyed or damaged as a result of other uses. Creation also occurs when someone wants to develop a wetland habitat at a specific site to attract birds or to perform other wetland functions such as water purification. It requires the establishment of a suitable water source and hydraulic regime that will support artificially seeded or planted vegetation. Successful developments are those where the water problem is resolved before the structure is built. In strip-mine areas, where we have studied plant response to areas virtually free of seed banks (owing to soil-handling techniques), natural seeding of common wetland annuals, sub-mergent vegetation, and shoreline willows requires from 5 to 10 years for establishment of the most minimal wetland plant assemblages; similar timing of plant and bird occupancy has also been noted in restored wetlands in Iowa (Van Rees-Siewert and Dinsmore 1996). Efforts to create small basins in Wisconsin showed very slow invasion of native plant, and seeding was used to enhance diversity within 3 years (Reinartz and Warne 1993). Some inverte-brates invade in a matter of weeks or months, and herptiles and fish are highly variable and dependent on connectivity to other wetlands (Weller 1995). These in turn dictate whether, how, and when they will be used by waterbirds.

Despite our good fortune in that many wetlands can be managed, restored, and even created, we should not be too complacent nor feel we can be liberal in destroying the natural and old in place of the new and the young. There is much we do not know, but it is certain that the complexity of the more sustained wetland system is great and that the pioneering successional stages are those we most readily duplicate. Birds are but one component of this complex system and are clearly successful because they are so adaptable. Insects and crustaceans may be equally so but many other animals are not. Moreover, we typically deal with a fragment of the long time scale or the vast geographic sphere of mobile waterbirds. From all that has been observed and measured, we assume that each wetland makes a difference, that each influences birds locally, and that often birds move long distances. Future studies will need to determine the reality of these and similar assumptions, but until we know more, conservation of all types of wetland in all places is the only safe strategy for the conservation of wetland birds. As in most things, what is good for natural systems is good for society; it just takes a little longer to understand and appreciate it!

15.2 Impact assessment and advanced planning

Driven by current environmental laws and concerns, one of the common efforts of conservation agencies and consulting firms is the use of ecological and biological knowledge to predict what impacts human-induced actions will have on habitat. This approach arose in the USA in relation to damming or modification of large river systems for flood control and alternative uses of water, such as irrigation, but similar efforts have been necessary over much of the world. However, the usual situation was not prediction but rather demonstration of the effects for legal purposes, such as a demand for mitigated replacements, and this was only possible in those rare cases where pre- and post-impact data were available. The meandering and unique Kissimmee River in central Florida is a classic example of a river channelized to minimize flooding (Blake 1980) during the period when environmental awareness was increasing. The result was an environmental impact assessment that predicted devastating effects on the abundance and diversity of wildlife and on human recreational use. However, the project had been popular and well-financed and was too far along to stop. Predictions were realized immediately because of valuable pre-channelization data on wetland animals (Perrin *et al.* 1982), and actions were set in motion to reverse the damage. Millions of dollars were spent in the modifications, and millions more will be spent in the restoration – one of the largest ever to be undertaken (Koebel 1995). Prediction of waterbird response to the return of seasonal flows and restoration of the diverse wetlands used a modified guild approach to estimate changes in species composition based on habitat diversity (Fig. 15.1).

Such approaches have the potential to prevent destruction of natural areas that provide multiple benefits to society, and recreation and human enjoyment of natural values carry more weight than they once did. In most cases, the greatest alarm has been registered where endangered or threatened species have been involved. Depending on the country, such legal protection has been responsible for saving not only breeding areas for the species but also important habitat for the entire animal community. However, species-oriented legislation may not clearly address the community or focus on the essential habitat. In the USA this resulted in the development of **Habitat Conservation Plans** to promote recognition and early resolution of an obvious and rising problem. Such plans have been developed where one or several species of concern are intertwined with major commercial developments or resource uses and allow advance negotiation and sometime mitigation to ensure that all activities can proceed. Obviously, this is not the same as total preservation of the species or community, and the quality of the plan is dependent upon the scientific underpinning of the requirements and the effectiveness of the negotiators (Luoma 1998). Moreover, because multiple habitats may be involved, some types of natural area (small wetlands) may be sacrificed to save those of greater importance for the species of focus. Realistic mitigation and replacement of these may be impossible, particularly when a bottomland is permanently flooded.

15.3 Population management

Because of the scientific advances in rearing of endangered wildlife and of releasing them into the wild, and the necessity of ensuring their subsequent survival, a few comments are required to link these efforts to habitat. Except for preservation for maintenance of the species as captives, it does little good to rear and release endangered species into the wild if there is no habitat remaining, if important mortality factors have not been corrected, or if environmental contaminants are still present. Protection of remaining habitat must be the first step, and while many biologists recognize this, only recently has the legal authority of the higher courts, in the USA at least, dictated that further damage to habitat must be prevented. Subsequently, protection and enhancement of habitat should still be the main strategy for survival of threatened species.

Direct, species-oriented management often includes the provision of nest sites for those species that respond to structure (see earlier comments), foods for those that feed in areas where management is feasible and can be done at reasonable cost, and predator control. Wildlife managers periodically have been embroiled in predator control to protect a selected species (especially game or endangered species), and although many species have shown positive responses, there seem to be few cases where actual populations of the target species have been increased (Cote and Sutherland 1997). Clearly, meeting the habitat needs also involves meeting the needs for cover and escape from predators.

Another population-management issue beyond the coverage of this book but interrelated with habitat and conservation issues is hunting. Major taxa legally hunted in most parts of the world include waterfowl, coots and other rails, cranes, and selected shorebirds such as snipe and woodcock. Hunting of other shorebirds and smaller wetlands birds has been prohibited in most parts of the world. While there is no doubt that hunting kills birds and can be detrimental to populations in some circumstances, most populations seem to be sufficiently productive that a high rate of annual mortality is expected and compensated for through lowered natural mortality or increased production. In most nations, harvests are tightly regulated as to time of the year and day, number of individual birds taken per hunter, per day, number of hunters per unit area, etc. The current approaches are reviewed in a number of works (Baldassare and Bolen 1994, Johnson, Nichols and Schwartz 1992) and generally have demonstrated that harvests can be regulated safely. Obviously, closure of seasons is one of the forms of legal management that can be dealt with promptly. Moreover, many hunting groups devote major effort to preservation of wetland habitat, which benefits all wetland wildlife.

References

Baldassare, G. A. and Bolen, E. G. (1994). *Waterfowl ecology and management.* New York: Wiley.

Blake, N. M. (1980). *Land into water – water into land.* Tallahassee, FL: University Presses of Florida.

Burger, J., Schisler, J., and Lesser, F. H. (1982). Avian utilisation on six salt marshes in New Jersey. *Biological Conservation* 23, 187–212.

Cote, I. M. and Sutherland, W. J. (1997). The effectiveness of removing predators to protect bird populations. *Conservation Biology* 11, 395–405.

Diamond, J. M. (1975). The island dilemma: lessons of modern biogeographic studies on the designs of natural preserves. *Biological Conservation* 7, 129–46.

Dobson, A. P., J. P. Rodriguez, W. M. Roberts, and Wilcove, D. S. (1997). Geographic distribution of endangered species in the United States. *Science* 275, 550–3.

Dugan, P. J. (Ed.). (1990). *Wetland conservation, a review of current issues and required action.* Gland, Switzerland: World Conservation Union, IUCN.

Fredrickson, L. H. and Reid, F. A. (1990). Impacts of hydrologic alteration on management of freshwater wetlands. In *Management of dynamic ecosystems*, ed. J. M. Sweeney, pp. 71–90. West Lafayette, IN. The Wildlife Society North-Central Section.

Fredrickson, L. H. and Taylor, T. S.(1982). Management of seasonally flooded impoundments for wildlife. pp. 1–29. Resource Publication No. 148. Washington, DC: U S Fish & Wildlife Service.

Holling, C. S. (ed.). (1978). *Adaptive environmental assessment and management.* New York: Wiley.

Johnson, D. H. and Igl, L. D. (1995). Contributions of the Conservation Reserve Program to populations of breeding birds in North Dakota. *Wilson Bulletin* 107, 709–18.

Johnson, D. H., Nichols, J. D., and Schwartz, M. D. (1992). Population dynamics of breeding waterfowl. In *Ecology and management of breeding waterfowl*, eds. B. D. J. Batt, A. D. Afton, M. G. Anderson, C. D. Ankney, D. H. Johnson, J. A. Kadlec, and G. L. Krapu, pp. 446–85. Minneapolis, MN: University of Minnesota Press.

Kiester, A. R., Scott, J. M., Csuti, B., Noss, R., Butterfield, B., Sahr, K., and White, D. (1996). Conservation prioritization using GAP data. *Conservation Biology* 10, 1332–42.

Koebel, J. W. Jr (1995). An historical perspective on the Kissimmee River restoration project. *Restoration Ecology* 3, 149–59.

Kushlan, J. A. (1979). Design and management of continental wildlife reserves: lessons from the Everglades. *Biological Conservation* 15, 281–90.

Luoma, J. R. (1998). Habitat conservation plans: compromise or capitulation? *Audubon* 100: 36–43.

MacArthur, R. H. and Wilson, E. O. (1967). *The theory of island biogeography.* Princeton, NJ: Princeton University Press.

National Fish and Wildlife Foundation (1997). *Directory of birding festivals, 1997.* Washington DC: National Fish and Wildlife Foundation.

Noss, R. F. and Cooperrider, A. (1994). *Saving nature's legacy; protecting and restoring biodiversity.* Washington, DC: Island Press.

Payne, N. F. (1992). *Techniques for wildlife habitat management of wetlands.* New York: McGraw-Hill.

Perrin, L. S., Allen, M. J., Prowse, L. A., Montalbano III, F., Foote, K. J., & Olinde, M. W. (1982). *A report on fish and wildlife studies in the Kissimmee River Basin and recommendations for restoration.* Okeechobee, FL: Florida Game and Fresh Water Fish Commission.

Reinartz, J. A. and Warne, E. L. (1993). Development of vegetation in small created wetlands in Southeastern Wisconsin. *Wetlands* 13, 153–64.

Riley, L. and Riley, W. (1992). *Guide to the National Wildlife Refuges.* New York: Macmillan.

Scott, D. A.(ed.) (1982). *Managing wetlands and their birds.* Slimbridge, UK: International Wildfowl Research Bureau.

Scott, J. M., Davis, F., Csuti, B., Noss, R., Butterfield, B., Groves, C., Anderson, H., Caicco, S., D'Erchia, F., Edwards T. C. Jr, Ulliman, J., and Wright, R. G. (1993). Gap analysis: a geographic approach to protection of biological diversity. *Wildlife Monographs* 123, 1–41.

Shafer, C. L. (1995). Values and shortcomings of small reserves. *BioScience* 45, 80–8.

Stanley, T. R. Jr (1996). Ecosystem, management and the arrogance of humanism. *Conservation Biology* 9, 255–62.

van der Valk, A. G. and Davis, C. B. (1978). The role of seed banks in the vegetation dynamics of prairie glacial marshes. *Ecology* 59, 322–35.

Van Rees-Siewert, K. L. and Dinsmore, J. J. (1996). Influence of wetland age on bird use of restored wetlands in Iowa. *Wetlands* 16, 577–82.

Walters, C. J. (1986). *Adaptive management of renewable resources.* New York: McGraw-Hill.

Weller, M. W. (1978). Management of freshwater marshes for wildlife. In *Freshwater wetlands, ecological processes and management potential,* eds. R. E. Good, D. F. Whigham, and R. L. Simpson, pp. 267–84. New York: Academic Press.

Weller, M. W. (1979). Birds of some Iowa wetlands in relation to concepts of faunal preservation. *Proceedings of the Iowa Academy of Science* 86, 81–8.

Weller, M. W. (1990). Waterfowl management techniques for wetland enhancement, restoration and creation useful in mitigation procedures. In *Wetland creation and restoration,* eds. J. A. Kusler and M. E. Kentula pp. 517–28. Washington DC: Island Press.

Weller, M. W. (1994). Seasonal dynamics of bird assemblages in a Texas estuarine wetland. *Journal of Field Ornithology* 65, 388–401.

Weller, M. W. (1995). Use of two waterbird guilds as evaluation tools for the Kissimmee River restoration. *Restoration Ecology* 3, 211–24.

Weller, M. W. (1996). Birds of rangeland wetlands. In *Rangeland Wildlife,* ed. P. Krausmann, pp. 71–82. Denver, CO: Society for Range Management.

16
Outlook

Those who have observed wetland birds and noted changes in wetland habitats over time share a great concern about whether these birds can and will survive. Obviously, they cannot survive without wetlands. Fortunately, wetlands are amazingly resilient systems because they are communities of diverse organisms adapted to dynamic water regimes; some one of the many possible species – whether algae, forb, invertebrate or bird – seems able to succeed when others do not, and the system seems to function at some biochemical and trophic level with these alternatives. Birds show even more flexibility than most members of the community because of their great mobility, but because they are dependent on water and food resources, they are no less sensitive to external impacts on the system. While some bird species usually can find resources for survival, fewer can fulfill the needs for successful reproduction in stressed wetlands, which results in a community with reduced species richness caused by reduced habitat diversity, poor conditions for food organisms, and reduced water quality.

We find amazing examples of wetland birds that live with society: Killdeer and Black Skimmers nesting on rooftops and feeding elsewhere; Ospreys, geese, storks, and herons nesting on various artificial structures and flying long distances for food; Brown Pelicans, Great Egrets, and Great Blue Herons gathering at boat docks to mooch food from incoming fishermen; White Pelicans feeding at night by pier lights; gulls feeding behind fishing boats; pipits, harriers, egrets, plovers, and gulls and terns feeding behind the plow or the still-burning fire in agricultural fields; and mass migrations of ducks, geese, swans, cranes, and pelicans along their traditional river routes that now take them directly over sprawling and spewing cities.

Many people have had the experience of visiting countries with human populations at levels that have seriously affected wetland habitats and bird resources. Yet, we are surprised at the numbers if not the variety of waterbirds seen. A few species use constructed wetlands designed for various alternative uses: ducks may find submergent vegetation or invertebrates in gravel pits (Svedarsky and Crawford 1982); many birds species feed and nest in rice fields (Fasola and Ruiz 1996), and similar agricultural fields are used in the non-breeding season; flamingos use salt-ponds (Espino-Barros and Baldassarre

1989); and ducks and shorebirds find rest and water in livestock ponds (Evans and Kerbs 1977), or food, cover, and even nest sites in any unused low-lying areas or those used for light grazing (e.g., Killdeer).

Wetlands now are built or modified as tertiary sewage treatment systems, producing enormous quantities of invertebrates favored by swimming water-birds that strain for their foods (Swanson 1977). Minnows are an adaptable and diverse group common to many types of wetland and attract bitterns, egrets, and herons. I have visited several such treatment units that have become favored bird observation sites, with an impressive array of birds and with wonderful opportunities to observe foraging behavior. Northern Shovelers and Ruddy Ducks find the enriched waters a great place for invertebrate foods; Tricolored Herons, several egrets, and the seemingly shy Least Bitterns and Clapper Rails also demonstrate their feeding tactics. Not all come to eat; Redheads and Northern Pintails come to drink, rest, and avoid the wind on open bays. At one such unit, Frigatebirds briefly scoop up fresh water and sail on. American Coots, Moorhens, and Pied-billed Grebes nest with little obvious concern for those who pass by.

In many cases, use of such areas is temporary and by migrants and there-fore, is the product of breeding successes elsewhere, which makes us wonder whether that source also will fade under future human stresses. But some populations breed successfully and maintain themselves for many years in areas that seem to us unsuitable, presumably because the essential habitat resources are there despite the seeming stresses of society (e.g., Ospreys, Bald Eagles, skimmers, terns, herons, and kites). Some have maintained free-living populations within major cities for many years, such as the breeding Tufted Ducks, Common Pochards, and Mallards of London parks (Gilham 1986), and Mallards of the northeastern USA (Figley and van Druff 1982). Others, like Mute Swans, have been domesticated and managed as decorative flocks for centuries (Ticehurst 1957). In some cases, however, we have succeeded too well because we, in effect, have created a predator-free habitat, a non-hunting zone, and excellent food or water supplies that may lead to overpopulations of attractive but often nuisance species like Canada Geese – now widely distrib-uted around the world (Cooper and Keefe 1997, Hanson 1965).

Addition of water areas for livestock or water supplies and associated agri-cultural crops often brings new species into an area, such as Maned Ducks in Australia (Kingsford 1992) and Black-bellied Whistling Ducks in south Texas (Bolen and Forsythe 1967, Bolen and Rhylander 1983). Other examples of the adaptability of selected species to large and permanent water are species like cormorants, which seem to have increased in areas where we have created ponds and lakes, and then we complain because they eat "our" fish! In the USA, the rate of wetland loss should be declining because of a declared policy of "no-net loss," but our wetland surveys show an increasing proportion of open-water types and do not detect degradation of wetlands, which tend to have lesser quality because of extremes in water volume and water quality, and reduced vegetation. In most cases, replacement of natural wetlands with artificial ones do not maintain the natural bird diversity of the area (Broome, and Jarman 1983, Erwin, Coulter and Coggswell 1986) because open water areas tend to replace dynamic, shallow, and richly vegetated wetlands, which have greater habitat diversity. Moreover, mitigated wetlands that are created or restored as substitutes for direct losses rarely are the equivalent in habitat value.

Despite these concerns, we have reason for optimism over progress in the period since the early 1980s – at least in wealthy countries with well-educated societies. Public interest in environmental issues seems steadily to increase, as indicated by political actions, private interest groups, and fund-raising efforts, and because of this some major successes have been achieved. Educational coverage of habitat conservation as well as species at risk is growing. Zoos, nature centers, and environmentally aware industries are enhancing, creating, and using wetlands as examples of exciting and bird-rich ecosystems that need protection. There is increased understanding that protecting endangered species requires first the protection of their habitat as well as preventing mortality or displacement to less productive habitats. Several species now are classed as recovered or recovering (North American Bald Eagle, several egrets, Peregrine Falcon, Whooping Crane), reflecting major national and international decisions and policies that have teamed government, industry, and the public in successful recovery programs (Cannon 1996).

These birds and other wetland values will persist if we develop a plan for society that includes recognition of and respect for these and other natural systems. In many cases, this does not mean we cannot share and use these wetland resources (Maltby 1986), but we must know and accept certain limitations to capitalize on other advantages. As knowledge, insight, and interest expand, we see some impressive wetland conservation results worldwide, but results demand cooperation of each area, whether city, county, state, or country level to establishment planning goals to avoid losing the local and small as well as the distant and large (Kusler 1983). It is the composite of these various natural and diverse habitat units that make possible the survival of species that use small habitat patches and those that use large and linear units; this is essential to sustain species richness and biodiversity.

In those parts of the world where human survival is the first consideration, and where standards of living are either decreasing (or increasing only at the sacrifice of stressing natural resources), further deterioration of natural values is inevitable. Birds are good indicators of these changes in wetlands but sometimes show deceptively large numbers of a few species. It is essential to have good measures of the health or ecological integrity of the system because the maintenance of these natural habitats not only perpetuates typical assemblages of birds for that region and wetland type but also dictates the survival of global migrants. Therefore, planning groups must recognize that wetland resource values go far beyond birds to include resources for many human cultures and functional values of the biosphere. Where such values are esthetic or recreational rather than monetary, wetlands have been preserved through outright purchase and lease, but is this feasible in third-world nations as human populations continue to rise (Errington 1996)? As Aldo Leopold recognized many years ago (Leopold 1935, 1949), conservation of remaining wildlife habitats and restoration of those recently lost probably is feasible only as individuals develop a vested interest and conservation ethic that recognizes resource values, have some access to the resources, and accept responsibility for them.

Certainly, pristine wetlands of original size, density, and characteristics would be an ideal goal and we should establish that goal for major wetland types in each region of each country or political unit. Such areas have been set aside to some degree in refuges and preserves, but many of these have traditional grazing and mineral rights and other land-uses. Moreover, many have modified water regimes outside the preserve boundaries that seriously influence how the "protected" system functions. Perhaps the only National Park in the USA that is totally a wetland system is Everglades National Park in Florida, but external uses and diversion of water has made that a highly impacted area made up of patches of modified wetlands that no longer function as a unit (see Douglas 1947). Recent state, local, and national agreements to repair the "plumbing" in this system and to reduce agricultural pollution hold real promise for this unique wetland. Natural preserves like Big Cypress in Florida and the Big Thicket of east Texas have major wetland components.

As the public becomes concerned and more knowledgeable about ecological approaches to conservation, there are dangers of focusing the public on the wrong goals and values and improper temporal and spatial scales to simplify and to gain support. This is especially true if guidelines are based on a pattern of terrestrial systems, which operate on a more annual time frame. On a still broader basis, the concept of biodiversity is very popular, and a variety of other richness-based methods of delineation are in use (e.g., Williams *et al.* 1996), but we must ensure that the identification of clear goals precedes development of rigid approaches and strategies. There are some frightening interpretations of richness data inferring that we can save many of the species of the world by acquiring only 2% of the land areas (US News and World Report 1997); this includes mostly tropical forests where species richness of breeding residents is high and tropical islands where endemic species tend to have low populations – and therefore, are threatened or endangered. Such a plan could rob society of a high percentage of the high-latitude wetlands and wetland birds of the world, and unique examples of habitat exploitation strategies by a single species that spans the extremes of both hemispheres! Concurrently, a strategy of this type would minimize protection of many unique wetland habitats dispersed elsewhere on the planet but which may not have high concentrations of endemic species and yet are productive and valuable resources with multiple functions serving humans as well as other diverse and unique animals (Maltby 1986).

Outlook

Therefore, biodiversity should not be interpreted as a species count but rather as a unique collection of organisms in a unique ecosystem setting that probably cannot be replicated and that cannot be moved to another site because of the environmental drivers. Some workers have recognized that habitat, community, and ecosystem quality are not easily measured, and that we tend to look for simple indices. As a result, some scientists encourage a focus on the maintenance of healthy and self-sustaining ecosystems, termed **biological integrity** (Angermeier and Karr 1994, Karr 1991), as opposed to biodiversity or "hotspots" alone.

References

Angermeier, P. L. and Karr, J. R. (1994). Biological integrity versus biological diversity as policy directives. *BioScience* 44, 690–7.

Bolen, E. G. and Forsythe, B. J. (1967). Foods of the black-bellied tree duck in south Texas. *Wilson Bulletin* 79, 43–9.

Bolen, E. G. and Rylander, M. K. (1983). *Whistling-ducks: zoogeography, ecology, anatomy.* Special Publication No. 20. Lubbock, TX: Texas Technical University Museum.

Broome, L. S. and Jarman, P. J. (1983). Waterbirds in natural and artificial waterbodies in the Noamoi Valley, New South Wales. *Emu* 83, 99–104.

Cannon, J. R. (1996). Whooping crane recovery: a case study in public and private cooperation in the conservation of endangered species. *Conservation Biology* 10, 813–21.

Cooper, J. A. and Keefe, T. (1997). Managing urban Canada Geese: policies and procedures. *Transactions of the North American Wildlife & Natural Resource Conference* 62, 412–30.

Douglas, M. S. (1947). *The everglades: river of grass.* St Simons, GA: Mockingbird Books.

Errington, P.L. (1996). *Of men and marshes.* Ames, IA: Iowa State University Press.

Erwin, R. M., Coulter, M., and Coggswell, H. (1986). The use of natural vs. man-modified wetlands by shorebirds and waterfowl. *Colonial Waterbirds* 9, 137–256.

Espino-Barros, R. and Baldassarre, G. A. (1989). Activity and habitat-use patterns of breeding Carribean Flamingos in Yucatan, Mexico. *Condor* 91, 585–91.

Evans, K. E. and Kerbs, R. R. (1977). *Avian use of livestock watering ponds in western South Dakota.* General Technical Report RM-35. Washington, DC: U S Forest Service.

Fasola, M. and Ruiz, X. (1996). The value of rice fields as substitutes for natural wetlands in the Mediterranean Region. *Colonial Waterbirds* 19, 122–8.

Figley, W. K. and van Druff, L. W. (1982). The ecology of urban mallards. *Wildlife Monographs* 81, 1–40.

Gilham, E. (1986). *Tufted Ducks in a Royal park.* Romney Marsh, Kent, UK: Gilham.

Hanson, H. C. (1965). *The Giant Canada Goose.* Carbondale, IL: Southern Illinois University Press.

Karr, J. R. (1991). Biological integrity: a long-neglected aspect of water resource management. *Ecological Applications* 1, 66–84.

Kingsford, R T. (1992). Maned ducks and farm dams: a success story. *Emu* 92, 163–9.

Kusler, J. A. (1983). *Our national wetland heritage, a protection guidebook.* Washington DC: Environmental Law Institute.

Leopold, A. (1935). Coon Valley: an adventure in cooperative conservation. *American Forests Magazine* 5, 1–4.

Leopold, A. (1949). *Sand County almanac.* Oxford: Oxford University Press.

Maltby, E. (1986). *Waterlogged wealth.* London: Earthscan, International Institute for Environment and Development.

Svedarsky, D. and Crawford, R. D. (eds.) (1982). *Wildlife Values of Gravel Pits.* Miscellaneous Publication 17–1982. St. Paul, MN: University of Minnesota Agricultural Experiment Station.

Swanson, G. A. (1977). Diel food selection by Anatinae on a waste-stabilization system. *Journal of Wildlife Management* 41, 226–31.

Ticehurst, N. F. (1957). *The Mute Swan in England.* London: Cleaver Hume.

US News and World Report. (1997). *Biodiversity; bang for the conservation buck.* US *News and World Report* February 28, 43.

Williams, P., Gibbons, D., Margules, C., Rebelo, A., Humphries, C., and Pressers, R. (1996). A comparison of richness hotspots, rarity hotspots, and complementary areas for conserving diversity of British birds. *Conservation Biology* 10, 155–74.

17

Epilogue

Wetlands are exciting places for many reasons, but birds are among the more prominent attractions. Birds are perhaps the most conspicuous component of a diverse biotic community with complex physical drivers interwoven in an ecosystem that is both unique and important. Wetlands have played an important role in supplying the resources needed for many societies and still function in that way for some groups today, but they also have provided an ideal setting for developing the scientific understanding of how such complex ecological systems function. But many wetlands are in jeopardy, and birds will be among the first indicators of dangers ahead for an individual wetland or for a wetland type or region. Despite the amazing resilience of many wetland types (which we can and do exploit to our benefit), others are delicate and probably irreplaceable. Moreover, they are sensitive to seemingly unrelated events both inside and outside the system that can change so much and so fast, and often to the detriment of waterbirds. Few habitats demonstrate the importance of one major driver, water, on the complexity, organization, and dynamics of an ecosystem. Although all habitats respond to water, wetlands have short-term and long-term consequences that dictate the biodiversity and trophic structure of the entire community – sometimes for hundreds of years – through plant and animal succession. Although temperature can also be very important, it is always secondary to the availability and timing of water.

It should be clear by now that habitat and its diverse resources determine in many direct and indirect ways the evolution of individual species and the relationships between species of a community. As we have seen, individual and species' adaptations include anatomy, morphology, physiology, and behavior. Assemblages involve interrelationships such as predation, competition, and diet specializations. While all birds respond to their habitat in this way, it is especially dramatic in wetland species because of the richness of the system, which creates dense populations, complex social behavior, diverse interspecies interactions, and close ties to the biological and physical influences of the habitat. Through their structural diversity and water regimes, wetland types dictate assemblages of birds (obviously influenced by geographic area), and these function as communities that may be skewed in species composition by the immediate characteristics of the vegetation and water. The uniqueness and integrity of such communities are reason for amazement, in part because they are ever changing and are remarkably successful in view of the dynamics of their habitat.

It is this uniqueness that makes us anxious to preserve more than relicts of these diverse wetland habitats. Without doubt, we can have simple wetland systems under stressed conditions that satisfy the needs of a small number of tolerant species living together as a community, but we can retain the complex systems of dynamic species only with the preservation of the environmental features that induce and maintain their integrity. When most water is intensively managed for alternative uses, when basins are filled, and when streams can no longer flood and scour, fewer and more simple wetland types will result, and lower diversity and less adaptive and resilient communities of birds can be expected. We will all be the poorer for it.

Part of our conservation problem is that we lack a clear-cut goal for wetlands. Moreover, because of the multiplicity of interests represented, agreement on such targets will always be a process of optimization, with the probability that the majority will be unhappy. However, the alternative of total elimination is far more serious. Not only must these individuals and peoples work together toward somewhat evasive goals, they also chase a moving target that is rarely understood by laypersons and politicians. Therefore, we often focus on the short term and small scale and rarely can achieve funding to pursue research or management efforts that are essential.

Few habitats demand more diversity of scientific expertise, or more integration of system functions and interactions than do wetlands. The wetlands survive only by cooperative efforts with political and socioeconomic entities, with businesses on the land and in the cities, and with primitive societies eking out a subsistence-level way of life. Wetlands demand the attention of wildlife biologists, plant ecologists, limnologists, stream ecologists, marine biologists, physical oceanographers, soil specialists, geologists, anthropologists, urban planners, city- to national-level politicians, and global economists and financiers.

Ecosystem-level research at this scale is more expensive than can be imagined and much less predictable than building a defense arsenal or funding research on a new food resource for starving people. It is understandably difficult for politicians to spend vast sums on the nebulous outcome of such research, but the ultimate outcome of ecosystem-level research has potential application far beyond birds or bird habitat.

Variability, instability, and unpredictability are all characteristics of most productive wetland habitats for wildlife. Nevertheless, it is important that we understand and learn to live with this variability. It does not make the conservation challenge any easier because it demands that sophisticated educational efforts precede group agreement and it makes it dangerous to make promises or encourage expectations about complex and variable systems. Although we are doing better at getting people to appreciate wetland values and functions, the magnitude of the problems facing conservationists seems to grow, and short-term economic gain is a difficult challenge to overcome when argued for the long-term resource values of functional ecosystems. But like the perspective of habitat, it is a concept that must be understood because it will influence the sustainability of life on earth.

Appendix 1 Scientific names of birds and bird groups

The scientific names of birds or bird groups mentioned in this book are listed alphabetically by English name. Most nomenclature is derived from Monroe and Sibley (1993), American Ornithologists' Union *Check-list of North American Birds* (1983) and its supplements to 1997, and regional works cited elsewhere.

Albatross (Diomedeidae)

Albatross, Black-browed *Diomedea melanophris*

Albatross, Grey-headed *Diomedea chrysostoma*

Albatross, Light-mantled Sooty *Phoebetria palpebrata*

Alseonax, Swamp *Muscicapa aquatica* (Africa)

Anhinga *Anhinga anhinga* (Americas)

Auk, Great *Pinguinus impennis* (extinct)

Avocet *Recurvirostra* spp. (most continents)

Avocet, American *Recurvirostra americana*

Babbler, Chestnut-capped *Timalia pileata*

Babbler, Marsh *Pellorneum palustre*

Bishop, Red *Euplectes orix*

Bittern, American *Botaurus lentiginosus*

Bittern, Least *Ixobrychus exilis*

Blackbirds, New World (Icteridae)

Blackbird, Brewer's *Euphagus cyanocephalus*

Blackbird, Red-winged *Agelaius phoeniceus*

Blackbird, Rusty *Euphagus carolinus*

Blackbird, Scarlet-headed *Amblyramphus holosericeus*

Blackbird, Tricolored *Agelaius tricolor*

Blackbird, Yellow-headed *Xanthocephalus xanthocephalus*

Blackbird, Yellow-winged *Agelaius thilius*

Bobolink *Dolichonyx oryzivorus*

Boobies (large marine plunge-divers related to pelicans)

Brant (American usage) or Brent (Eurasian usage) *Branta bernicla*

Brant, Black or Pacific Black *Branta bernicla nigricans*

Brant, White-bellied *or* Atlantic *Branta b. bernicla*

Bufflehead *Bucephala albeola*

Canary, Papyrus *Serinus koliensis*

Canvasback *Aythya valisineria*

Caracara, Chimango *Milvago chimango*

Catbird, Grey *Dumetella carolinensis*

Chickadee, Black-capped *Parus atricapillus*

Chickadee, Boreal *Parus hudsonicus*
Cinclodes or Shaketails *Cinclodes* spp.
(Furnariidae)
Cisticola, Carruther's *Cisticola carruthersi*
Coot, American *Fulica americana*
Coot, Common or European *Fulica atra*
Coot, Giant *Fulica gigantea*
Coot, Horned *Fulica cornuta*
Cormorants, Double-crested *Phalacrocorax auritus*
Cormorant, Flightless or Galapagos *Phalacrocorax harrisi*
Cormorants, Neotropic *Phalacrocorax brasilianus*
Coscoroba Swan, *Cygnus coscoroba*
Coucal, Pheasant or Swamp Cuckoo *Centropus phasianinus*
Coursers and pratincoles (Glareolidae)
Crab-plover *Dromas ardeola*
Crake *Porzana* spp., and other genera (Rallidae)
Crane, Common *Grus grus*
Crane, Sandhill *Grus canadensis*
Crane, Whooping *Grus americana*
Crocodile-bird, or Egyptian Plover *Pluvianus aegyptius*
Crossbill, White-winged *Loxia leucoptera*
Crow, Fish *Corvus ossifragus*
Cuckoo, Mangrove *Coccyzus minor*
Cuckoo, Swamp (or Pheasant Coucal) *Centropus phasianus*
Cuckoo, Yellow-billed *Coccyzus americanus*
Curlew, Bristle-thighed *Numenius tahitiensis*
Curlew, Eskimo *Numenius borealis*
Curlew, Eurasian *Numenius arquata*
Curlew, Long-billed *Numenius americanus*

Darter *Anhinga* spp.
Dipper, American *Cinclus mexicanus*

Dipper, Eurasian or White-throated *Cinclus cinclus*
Doves (Various ground birds of family Columbidae)
Duck, African Black *Anas sparsa*
Duck, American Black *Anas rubripes*
Duck, bay, (inland diving or pochards) (tribe Aythyini)
Duck, Black-bellied Whistling *Dendrocygna autumnalis*
Duck, Black-headed *Heteronetta atricapilla*
Duck, Blue *Hymenolaimus malacorhynchus*
Duck, Brown *Anas aucklandica*
Duck, dabbling *Anas* spp.
Duck, Fulvous Whistling *Dendrocygna bicolor*
Duck, Harlequin *Histrionicus histrionicus*
Duck, Hawaiian *Anas wyviliana*
Duck, inland diving (bay, or Pochards) (tribe Aythyini)
Duck, Labrador *Camptorhynchus labradorius*
Duck, Laysan *Anas laysanensis*
Duck, Long-tailed or Oldsquaw *Clangula hyemalis*
Duck, Mallard *Anas platyrhynchos*
Duck, Mandarin *Aix galericulata*
Duck, Maned *Chenonetta jubata*
Duck, Masked *Nomonyx dominica*
Duck, Mottled *Anas fulvigula*
Duck, New Zealand Grey *Anas superciliosa*
Duck, North American Wood *Aix sponsa*
Duck, Pacific Black *Anas superciliosa*
Duck (Teal) Red-billed *Anas erythroryhncha*
Duck, Redhead *Aythya americana*
Duck, Ring-necked *Aythya collaris*
Duck, Ruddy *Oxyura jamaicensis*
Duck, sea (scoters, eiders, mergansers, etc., tribe Mergini)

Duck, Spectacled *Anas specularis*
Duck, Steamer *Tachyeres* spp.(only one of four can fly well)
Duck, Torrent *Merganetta armata*
Duck, Tufted *Aythya fuligula*
Duck, White-backed *Thalassornis leuconotus*
Duck, White-headed *Oxyura leucocephala*
Duck, Wood *Aix sponsa*

Eagle, African Fish *Haliaeetus vocifer*
Eagle, Bald *Haliaeetus leucocephalus*
Eagle, Sea *Haliaeetus* spp.
Eagle, Steller's Sea- *Haliaeetus pelagicus*
Egret (Heron), Black *Egretta ardesiaca*
Egret, Cattle *Bubulcus ibis*
Egret, Great *Ardea alba*
Egret, Reddish *Egretta rufescens*
Egret, Snowy *Egretta thula*
Eider, Common *Somateria mollissima*
Eider, King *Somateria spectabalis*
Eider, Spectacled *Somateria fischeri*
Eider, Steller's *Polysticta stellari*

Falcon, Peregrine *Falco peregrinus*
Finches (various seed-eating passer-ines of several families)
Finfoot, African *Podica sengalensis*
Finfoot, American (Sungrebe) *Heliornis fulica*
Finfoot, Masked *Heliopais personata*
Flamingo, Greater (including Caribbean) *Phoenicopterus ruber*
Flamingo, Lesser *Phoenicopterus minor*
Flycatcher, Acadian *Empidonax virescens*
Flycatcher, Alder *Empidonax alnorum*
Flycatcher, Olive-sided *Contopus borealis*
Flycatcher, Swamp *Muscicapa aquatica*
Flycatcher, Willow *Empidonax traillii*
Flycatcher, Yellow-bellied *Empidonax flaviventris*
Frigatebird *Fregata* spp.

Gadwall *Anas strepera*
Gallinule, Purple *Porphyrio martinicus*
Gannet *Morus* (*Sula*) spp.
Gerygone (Warbler), Fairy *Gerygone palpebrosa*
Gerygone (Warbler), Mangrove *Gerygone levigaster*
Godwit, Hudsonian *Limosa haemastica*
Goldeneye, Common *Bucephala clangula*
Gonolek, Papyrus *Laniarius mufumbriri*
Goose, African Pygmy *Nettapus auritus*
Goose, Andean *Chloephaga melanoptera*
Goose, Barnacle *Branta leucopsis*
Goose, Canada *Branta canadensis* and *Branta canadensis maxima*
Goose, Egyptian *Alopochen aegyptiacus*
Goose, Greater White-fronted *Anser albifrons*
Goose, Greylag *Anser anser*
Goose, Kelp *Chloephaga hybrida*
Goose, Magpie (Pied) *Anseranas semipalmata*
Goose, Orinoco *Neochen jubata*
Goose, Snow *Anser caerulescens*
Goose, Spur-winged *Plectropterus gambensis*
Goose, Upland *Chloephaga picta*
Grackle, Boat-tailed *Quiscalus major*
Grackle, Common *Quiscalus quiscula*
Grebe, Eared *Podiceps nigricollis*
Grebe, Great-crested *Podiceps cristatus*
Grebe, Horned *Podiceps auritus*
Grebe, Pied-billed *Podilymbus podiceps*
Grebe, Red-necked *Podiceps grisegena*
Grebe, Western *Aechmophorus occidentalis*
Grouse, Ruffed *Bonasa umbellus*
Grouse, Spruce *Dendragapus canadensis*
Gull, Brown-hooded *Larus maculipennis*

Gull, Common Black-headed *Larus ridibundus*
Gull, Franklin's *Larus pipixcan*
Gull, Grey-headed *Larus cirrocephalus*
Gull, Herring *Larus argentatus*
Gull, Laughing *Larus atricapilla*
Gull, Ring-billed *Larus delawarensis*

Hamerkop or Hammerhead Stork *Scopus umbretta*
Harrier, Northern *Circus cyaneus*
Hawk, Black-collared *Busarellus nigricollis*
Hawk, Common Black *Buteogallus anthracinus*
Hawk, Red-shouldered *Buteo lineatus*
Heron (Egret), Black *Egretta ardesiaca*
Heron, Black-crowned Night- *Nycticorax nycticorax*
Heron, Great Blue *Ardea herodius*
Heron, Great White *Ardea herodius* (a white morph)
Heron, Green *Butorides virescens*
Heron, Grey *Ardea cinerea*
Heron, Little Blue *Egretta caerulea*
Heron, Tricolored *Egretta tricolor*
Heron, Yellow-crowned Night- *Nycticorax violacea*
Hoatzin *Opisthocomus hoatzin*

Ibis, Glossy *Plegadis falcinellus*
Ibis, Puna *Plegadis ridgwayi*
Ibis, White *Eudocimus albus*
Ibis, White-faced *Plegadis chihi*

Jacana (family Jacanidae)
Jaeger *Stercorarius* spp.
Jay, Blue *Cyanocitta cristata*
Jay, Grey *Perisoreus canadensis*

Killdeer *Charadrius vociferus*
Kingbird, Eastern *Tyrannus tyrannus*
Kingfisher, Amazon *Chloroceryle amazona*
Kingfisher, American Pygmy *Chloroceryle aenea*

Kingfisher, Green *Chloroceryle americana*
Kingfisher, Green-and-rufous *Chloroceryle inda*
Kingfisher, Mangrove *Halcyon sengaloides*
Kingfisher, Pied *Ceryle rudis*
Kingfisher, Ringed *Megaceryle torquata*
Kingfisher (Kookaburra), Shovel-billed *Clytoceyx rex*
Kiskadee, Great *Pitangus sulfuratus*
Kiskadee, Lesser *Pitangus lictor*
Kite, American Swallow-tailed *Elanoides forticatus*
Kite, Brahminy *Haliastur indus*
Kite, Hook-billed *Chondrohierax uncinatus*
Kite, Slender-billed *Rostramus hamatus*
Kite, Snail or Everglade *Rostramus sociablis*
Knot, Red *Calidris canutus*
Kookaburra, Laughing *Dacelo novaeguineae*
Kookaburra (Kingfisher), Shovel-billed *Clytoceyx rex*

Lapwing *Vanellus* spp.
Limpkin *Aramus guarauna*
Loon, Common *Gavia immer*
Loon, Pacific *Gavia pacifica*
Loon, Red-throated *Gavia stellata*

Mallard *Anas platyrhynchos*
Marshbird, Brown-and-yellow *Pseudoleistes virescens*
Marsh-Tyrant, White-headed *Arundinicola leucocephala*
Martin, Sand (Bank Swallow) *Riparia riparia*
Meadowlark *Sturnella* spp.
Merganser, Common *Mergus merganser*
Merganser, Hooded *Lophodytes cucullatus*
Merganser, Red-breasted *Mergus serrator*

Moorhen, Common *Gallinula chloropus*

Murrelet, Marbled *Brachyramphus marmoratus*

Nettapus (Goose, African Pygmy) *Nettapus auritus*

Night-heron, Black-crowned *Nycticorax nycticorax*

Night-heron, Yellow-crowned *Nycticorax violacea*

Notornis (Takahe) *Porphyrio manetelli*

Oldsquaw (Long-tailed Duck) *Clangula hyemalis*

Oriole, Northern *Icterus galbula*

Osprey *Pandion haliaetus*

Ovenbird (Neartic warbler) *Seiurus aurocapillus*

Ovenbirds *Cinclodes* spp. (Furnariidae of South America)

Owl, Barred *Strix varia*

Owl, Fish *Ketupa* spp.

Owl, Fishing *Scotopelia* spp.

Owl, Grass *Tyto capensis*

Owl, Great Horned *Bubo virginianus*

Owl, Marsh *Asio capensis*

Owl, Short-eared *Asio flammeus*

Oxpecker, Red-billed *Buphagus erythrorynchos*

Oystercatcher *Haematopus* spp.

Painted-Snipe *Rostratula* spp.

Parakeet, Monk *Myiopsitta monarchus*

Peeps (common name for a group of small sandpipers) *Calidris* spp.

Peewee, Eastern Wood *Contopus virens*

Pelican, American White *Pelecanus erythrorhynchos*

Pelican, Brown *Pelecanus occidentalis*

Pelican, Great White *Pelecanus onocrotalus*

Penguin, Gentoo *Pygoscelis papua*

Penguin, King *Aptenodytes patagonicus*

Petrels *Pterodroma* spp.

Petrel, Antarctic Giant *Macronectes giganteus*

Petrel, White-chinned *Procellaria aequinoctialis*

Phalarope, Red (Grey) *Phalaropis fulicaria*

Phalarope, Red-necked (Northern) *Phalaropis lobatus*

Phalarope, Wilson's *Steganopus tricolor*

Pheasant, Ring-necked *Phasianus colchicus*

Pigeon, Passenger *Ectopistes migratorius*

Pigeon, Rock *Columba livia*

Pintail, Northern *Anas acuta*

Pintail, South Georgia *Anas georgica georgica*

Pipit, American *Anthus rubescens*

Pipit, Rock *Anthus petrosus*

Pipit, South Georgia *Anthus antarcticus*

Pipit, Water *Anthus spinoleta*

Plover, American (Lesser Golden) *Pluvialis dominica*

Plover, Black-bellied (Grey) *Pluvialis squatarola*

Plover, Crab- *Dromas ardeola*

Plover, Egyptian (Crocodile-bird) *Pluvianus aegyptius*

Plover, Eurasian (Greater Golden) *Pluvialis apricaria*

Plover, Grey (Black-bellied) *Pluvialis squatarola*

Plover, Piping *Charadrius melodus*

Plover, Semipalmated *Charadrius semipalmatus*

Pochard, Common *Aythya ferina*

Pochard, African (Southern) *Aythya erythrophthalma*

Pochards (inland diving or bay ducks) (tribe Aythyini)

Pratincole, Collared *Glareola pratincola*

Pratincoles and coursers (Glareolidae)

Prion, Antarctic *Pachyptila desolata*

Quail (common name for small, ground-feeding galliforms)

Rail, Clapper *Rallus, longirostris*

Rail, King *Rallus elegans*
Rail, Sora *Porzana carolina*
Rail, Virgina *Rallus limicola*
Redhead *Aythya americana*
Redstart, American *Setophaga ruticilla*
Reedhaunter *Limnornis* spp.
Ruff *Philomachus pugnax*
Rushbird, Wren-like *Phleocryptes melanops*
Rush-Tyrant, Many-colored *Tachuris rubrigastra*

Sanderling *Calidris alba*
Sandpiper, Purple *Calidris maritima*
Sandpiper, Rock *Calidris ptilocnemis*
Sandpiper, Semipalmated *Calidris pusilla*
Sandpiper, Spotted *Tringa macularia*
Sandpiper, Western *Calidris mauri*
Sandpiper, White-rumped *Calidris fusicollis*
Scaup, Greater *Aythya marila*
Scaup, Lesser *Aythya affinis*
Scoter (sea ducks) *Melanitta* spp.
Scoter, Common (Black) *Melanitta nigra*
Scoter, Velvet (White-winged) *Melanitta fusca fusca*
Screamer, Southern (Crested) *Chauna torquata*
Sea-eagle, Steller's *Haliaeetus pelagicus*
Shag (term used for smaller cormorants of high latitudes)
Shelduck, Common *Tadorna tadorna*
Shelduck, Paradise (New Zealand) *Tadorna variegata*
Sheldgoose, Ashy-headed *Chloephaga poliocephala*
Sheldgoose, Egyptian *Alopochen aegyptiacus*
Sheldgoose, Upland *Chloephaga picta*
Shoebill or Whalehead *Balaeniceps rex* (heron-like, monotypic)

Shoveler (several species of shovel-billed ducks) *Anas* spp.
Shoveler, Northern *Anas clypeata*
Skimmer, Black *Rynchops niger*
Skua *Catharacta* spp.
Snipe, Common (Wilson's) *Gallinago gallinago*
Sora *Porzana carolina*
Sparrow, Henslow's *Ammodrammus henslowii*
Sparrow, House *Passer domesticus*
Sparrow, Le Conte's *Ammodrammus leconteii*
Sparrow, Lincoln's *Melospiza lincolnii*
Sparrow, Nelson's Sharp-tailed *Ammodrammus nelsoni*
Sparrow, Saltmarsh Sharp-tailed *Ammodrammus caudacutus*
Sparrow, Savannah *Passerculus sanwichensis*
Sparrow, Seaside *Ammodrammus maritimus*
Sparrow, Song *Melospiza melodia*
Sparrow, Swamp *Melospiza georgiana*
Spoonbill, Eurasian *Platalea leucorodia*
Spoonbill, Roseate *Ajaia ajaja*
Starling, Common *Sturnus vulgaris*
Starling, Slender-billed *Onychognathus tenuirostris*
Steamer-duck *Tachyeres* spp.
Stifftail (Ducks) *Oxyura* spp. (Oxyurini)
Stilt, Black-necked *Himantopus mexicanus*
Stone-curlew, Beach *Burhinus giganteus*
Stork, Hammerhead (Hamerkop) *Scopus umbretta*
Stork, Marabou *Leptoptiles crumeniferus*
Stork, White *Ciconia ciconia*
Stork, Wood *Mycteria americana*
Streamcreeper, Sharp-tailed *Lochmias nematura*
Sunbittern *Eurypyga serpentina*

Sungrebe (American Finfoot) *Heliornis fulica*
Surfbird *Aphriza virgata*
Swallow, Bank (Sand Martin) *Riparia riparia*
Swallow, Mangrove *Tachycineta albilinea*
Swallow, Rough-winged *Stelgidopteryx serripennis*
Swallow, Tree *Tachycineta bicolor*
Swallows (Hirundinidae)
Swallow-plover (pratincoles) (Glareolidae)
Swan, Black *Cygnus atratus*
Swan, Black-necked *Cygnus melanocorypha*
Swan, Coscoroba *Coscoroba coscoroba*
Swan, Mute *Cygnus olor*
Swan, Trumpeter *Cygnus buccinator*
Swan, Whooper *Olor cygnus*

Takahe (Notornis) *Porphyrio manetelli*
Tchagra, Marsh *Tchagra minuta*
Teal, Auckland Island Flightless *Anas aucklandica*
Teal, Blue-winged *Anas discors*
Teal, Brown *Anas aucklandica*
Teal, Green-winged *Anas crecca*
Teal, Laysan *Anas laysanensis*
Teal (Duck), Red-billed *Anas erythroryhncha*
Teal, Speckled *Anas flavirostris*
Tern, Antarctic *Sterna vittata*
Tern, Arctic *Sterna paradisaea*
Tern, Black *Chlidonias niger*
Tern, Common *Sterna hirundo*
Tern, Forster's *Sterna forsteri*
Tern, Least *Sterna antillarum*
Tern, Royal *Sterna maxima*
Tern, Sandwich *Sterna sandvicensis*
Tern, Whiskered *Chlidonias hybridus*
Thick-knee, Beach (Stone-curlew) *Burhinus giganteus*
Thick-knee, Water *Burhinus vermiculatus*

Thrush, Swainson's *Catharus ustulatus*
Tit, Marsh *Parus palustris*
Turnstone, Ruddy *Arenaria interpres*
Tyrant flycatchers (Tyrannidae)
Tyrant, Rush-, Many-colored *Tachuris rubrigastra*
Tyrant, Water- *Fluvicola* spp.
Tyrant, White-headed Marsh- *Arundinicola leucocephala*

Veery *Catharus fuscescens*
Vireo, Bell's *Vireo bellii*
Vireo, Mangrove *Vireo pallens*
Vireo, Red-eyed *Vireo olivaceous*
Vireo, White-eyed *Vireo griseus*

Wagtail, Cape *Motacilla capensis*
Wagtail, Grey *Motacilla cinerea*
Wagtail, Mountain *Motacilla clara*
Wagtail, Yellow *Motacilla flava*
Warbler, Aquatic *Acrocephalus paludicola*
Warbler, Arctic *Phylloscopus borealis*
Warbler, Bachman's *Vermivora bachmani*
Warbler, Black-throated Green *Dendroica virens*
Warbler (Gerygone), Fairy *Gerygone palpebrosa*
Warbler, Greater Swamp *Acrocephalus rufescens*
Warbler, Hooded *Wilsonia citrina*
Warbler, Lesser Swamp *Acrocephalus gracilirostris*
Warbler, Little Rush *Bradypterus baboecala*
Warbler (Gerygone), Mangrove *Gerygone levigaster* (Australia)
Warbler, Marsh *Acrocephalus palustris*
Warbler, Northern Parula *Parula americana*
Warbler, Paddyfield *Acrocephalus agricola*

Warbler, Palm *Dendroica palmarum*

Warbler, Prothonotary *Protonotaria citrea*

Warbler, Reed *Acrocephalus* spp.

Warbler, River *Locustella fluviatilis*

Warbler, Sedge *Acrocephalus schoenobaenus*

Warbler, Swainson's *Limnothlypis swainsonii*

Warbler, Yellow (includes N.A. Mangrove) *Dendroica petechia*

Warbler, Yellow-throated *Dendroica dominica*

Waterthrush *Seiurus* spp.

Whalehead (Shoebill) (heron-like, monotypic) *Balaeniceps rex*

Widowbird, Fan-tailed *Euplectes axillaris*

Widowbird, Marsh *Euplectes hartlaubi*

Wigeon, American *Anas americana*

Wigeon, Chiloe *Anas sibilatrix*

Wigeon, Eurasian *Anas penelope*

Willet *Catoptrophorus semipalmatus*

Woodcock, American *Scolopax minor*

Woodpecker, Ivory-billed *Campephilus princepalis* (extinct?)

Woodpecker, Red-bellied *Melanerpes carolinus*

Wren, Marsh *Cistothorus palustris*

Wren, Sedge *Cistothorus platensis*

Yellowlegs, Greater *Tringa melanoleuca*

Yellowlegs, Lesser *Tringa flavipes*

Yellowthroat, Common *Geothlypis trichas*

References

American Ornithologists' Union (1983). *Check-list of North American birds.* Washington, DC: American Ornithologists' Union.

American Ornithologists' Union (1997). Forty-first supplement to the American Ornithologists' check-list of North American birds. *Auk* 114, 542–52.

Monroe, B. L. Jr and Sibley, C. G. (1993). *A world checklist of birds*, 6th edn. New Haven, CT: Yale University Press.

Appendix 2 Scientific names of animals and animal groups other than birds

The scientific names of animal species or groups other than birds mentioned in this book are listed alphabetically by English name.

Alligator *Alligator mississippiensis*

Amphipoda: scuds and sideswimmers (Crustacea)

Arachnida: spiders and relatives

Barbus – (a genus of common African fish) (Cyprinidae)

Beaver *Castor canadensis*

Bobcat *Lynx rufous*

Buffalo, Water *Syncerus caffer*

Caddisfly (Tricoptera)

Carabidae (a family of common beetles)

Carp *Cyprinus carpio*

Cattle, Domestic *Bos taurus*

Chironomid, Midge (Chironomidae)

Cladocera: water fleas, e.g., Daphnia

Clam, Fingernail *Sphaerium* spp.

Clarius (African air-breathing catfish) *Clarius gariepinus*

Copepod, e.g., *Cyclops* spp. (Copepoda)

Coyote *Canis latrans*

Crab, Horseshoe *Limulus polyphemus*

Crayfish (Decapoda)

Crocodile, Nile *Crocodilus niloticus*

Cyclops (Copepoda)

Damselfly (Odonata)

Daphnia or water fleas (e.g., *Daphnia* spp. Cladocera)

Deer, Marsh *Cervus* spp.

Dog, Wild *Canis* spp.

Dragonflies (Odonata)

Elephant, African *Loxodonta africana*

Ephemeroptera (Mayfly)

Fly, Crane (Tipulidae)

Fox, Red *Vulpes fulva*

Gambusia (Mosquito-fish)

Hepsetus odoe (a predatory fish of Africa) (Hepsetidae)

Hippopotamus *Hippoppotamus amphibias*

Isopoda (fresh and marine aquatic sow-bugs) (Crustacea)

Labeo spp.: genus of a common African Cyprinid

Lechwe (an African antelope) *Kobus lechwe*

Mayfly (Ephemeroptera)

Midge (Chironomidae)

Mink *Mustela vision*

Minnows (Cyprinidae and other small fish)

Moose *Alces americana*

Mosquito (Culicidae)

Mosquito-fish *Gambusia* spp.
Muskrat *Ondatra zibethicus*
Mussel, e.g., *Mytilus* spp.: clinging
    marine mussels
Mussel, Zebra *Dreissena polymorpha*
Mytilus spp. (common genus of
    clinging marine mussels)

Nutria *Myocastor coypus*

Otter, River *Lutra canadensis*
Otter, Southern River *Lutra longicaudis*
Oyster *Crassostrea* spp.

Physidae (a common family of
    freshwater snails)
Platypus *Ornithorhynchus anatinus*

Rat, Rice *Oyzomys palustris*

Salamander, Tiger *Amhystoma
    tigrinum*
Scuds (sideswimmers) (Amphipoda)

Seal, Elephant *Mirounga leonina*
Shrimp, Fairy *Branchinecta* spp.
Shrimp, Seed (Ostracoda)
Sideswimmers (scuds)
    (Amphipoda)
Snail, Apple *Pomacea* spp.
Sowbugs (Isopoda)
Staphylinidae (a family of common
    beetles)
Stickleback, Brook *Culaea inconstans*
Stonefly (Plecoptera)
Swine, European Boar, (Feral Hog) *Sus
    scrofa*

Tilapia (a genus of common African
    fish) (Cichlidae)
Tipulidae (Crane Fly)
Tricoptera (Caddisfly)
Turtle, Snapping *Chelydra serpentina*

Water fleas *Daphnia* spp. (Cladocera)
Weasels *Mustela* spp.
Whirligig beetle (Gyrinidae)

Appendix 3 Scientific names of plants and plant groups

The scientific names of plants or plant groups mentioned in this book are listed alphabetically by English name or genus when used as a common name.

Algae *Spirulina* spp.
Arrowhead *Sagittaria* spp.
Azalea, Swamp *Rhododendron viscosum*

Baccharis (Groundsel-tree/bush) *Baccharis* spp.
Beggarstick *Bidens* spp.
Bladderwort *Utricularia* spp.
Blueberry, Highbush *Vaccinium corymbosum*
Bog-laurel *Kalmia polifolia*
Bulrush, Alkali *Scirpus paludosus*
Bulrush, Hardstem *Scirpus acutus*
Bulrush, River *Scirpus fluviatilis*
Bulrush, Softstem *Scirpus validus*
Burreed *Sparganium* spp.
Buttonbush *Cephalanthis occidentalis*

Cane, Giant *Arundinaria gigantea*
Cattail *Typha* spp.
Chufa (Flatsedge) *Cyperus esculentus*
Cottonwood *Populus deltoides*
Cranberry, American *Vaccinium macrocarpon*
Cypress, Bald- *Taxodium distichum*

Decodon (Swamp Loosestrife) *Decodon verticillatus*
Duckweed, Greater *Lemna major*
Duckweed, Lesser *Lemna minor*

Duckweed, Star *Lemna trisulca*

Elm, Water *Planera aquatica*

Fern, Cinnamon (Osmunda) *Osmunda cinnamonea*
Fir, Balsam *Abies balsamea*
Flatsedge (Chufa) *Cyperus esculentus*
Fleabane *Erigeron* spp.

Groundsel-tree/bush *Baccharis* spp.

Hydrilla *Hydrilla verticillata*

Iva (Marsh Elder) *Iva frutescens*

Jewelweed *Impatiens* spp.
Juncus (Rush) *Juncus* spp.

Kelp, Bull *Durvillea anarctica*
Kelp, Giant (Leafy) *Macrocystis pyrifera*

Leatherleaf *Chamaedaphne calyculata*
Lettuce, Sea *Lactuca* spp.
Lotus, American (Yellow) *Nelumbo lutea*

Mangrove (tropical seashore trees of many genera and families)
Mangrove, Red *Rhizophora mangle*
Maple, Red *Acer rubrum*

Marsh Elder (Iva) *Iva frutescens*
Millet (Barnyardgrass) *Echinochloa* spp.
Mountain-holly *Nemopanthus mucronata*

Nemopanthus (Mountain-Holly) *Nemopanthus mucronata*

Oak, Red *Quercus rubra*
Osmunda (Cinnamon Fern) *Osmunda cinnamonea*

Pitcher-plant, e.g., *Sarracenia* spp.
Plumegrass *Erianthus* spp.
Pondweed *Potamogeton* spp.
Pondweed, Sago *Potamogeton pectinatus*
Potato, Duck *Sagittaria* spp.

Reed, Common *Phragmites communis*
Rhodora *Rhodora canadense*
Rush *Juncus* spp.

Sedge *Carex* spp.
Smartweeds *Polygonum* spp.
Sphagnum Moss *Sphagnum palustre* and *S. rubrum*
Spikerush *Eleocharis* spp.
Spruce, Black *Picea mariana*
Sumac, Poison *Toxicodendron vernix*
Sundew *Drosera rotundifolia*

Tamarack (Larch) *Larix laricina*
Taro *Colocasia esculenta*
Tea, Labrador *Ledum groenlandicum*

Water-hyacinth *Eichhhornia grassipes*
Water-lily, Blue *Nymphaea elegans*
Water-lily, Pond *Nymphaea odorata*
Water-lily, Yellow *Nuphar advena*
Water-milfoil *Myriophyllum* spp.
Water-shield *Brasenia* spp.
Watermeal *Wolfia* spp.
Wigeongrass *Ruppia maritima*
Wild-celery, *Vallisneria americana*
Willow *Salix* spp.

Index of birds and bird groups

Index of birds and birds groups

Subject index

Upper-case roman numbers denote plates;
lower-case roman numbers denote Preface.

Adamus, 199, 207, 208
adaptations, 23, 27, 32, 43, 53, 74, 76, 108, 121, 123
adaptive management, 231
Africa, II*n*, 17, 24–7, 30, 32, 37, 41, 42, 73, 91, 114,
    213, 214–16
age, 63, 73, 74, 140
agencies, 133, 159, 193, 199, 203, 208, 216, 227,
    228, 233
aggression, 3, 5, 102, 113
agonistic, 99, 110
agriculture, I*d*, 37, 173, 193, 213, 214, 217, 221
Alaska, IV*k*, 19, 88, 134, 155, 174
algae, 33, 34, 67, 69, 94, 183, 184, 239
alligators, 19, 57
alpine, I*k*, 33, 53, 136
altricial, 30, 31, 35, 36, 42, 108
anatomy, 3, 49, 74, 245
aquaculture, 91, 173
Argentina, II*d*, 40, 66, 89, 91, 113, 157, 220
assemblage, 2, 8, 75, 77, 101, 153, 189, 192, 208, 217
assessment, 7, 121, 128, 200–3, 208, 233, 234
Auckland Islands, III*c*, V*a*
Australia, 24–7, 30, 32, 33, 87, 91, 157, 158, 213,
    220, 240

backwater, 17, 20, 53, 123, 139, 159, 160, 205, 232
Bald-cypress, 123
basin wetlands, I*a,k*, 14, 41, 51, 53, 123, 136, 138,
    148, 149, 151, 153, 155, 156, 159, 161, 167, 175,
    193, 196, 217
bathing, 49, 50, 55
Bay of Fundy, 20, 86, 205
beaters, II*n*, 71
beaver, 19, 133, 160, 167, 231
behavior
    innate, 6, 7, 49, 67
    learned, 6–8, 213, 228
Big Cypress, 242
Big Thicket, 242

biodiversity, xi, xii, 220, 221, 230, 241–3, 245
biogeographic, 2, 128
biogeography, 229
biological integrity, 243
biome, 2
biosphere, 242
biotic change, 145
biotic community, 245
biotope, 5
bird
    assemblage, 192, 217
    community, 2, 8, 70, 128, 228
    distribution, 23, 103, 124
    species diversity, 165, 173
    species richness, 51, 126, 134, 194, 206
bobcat, 58
body form, 33, 34
body size, 35, 41, 73, 92, 99, 100, 107, 109, 110
bog, 14, 30, 37, 38, 125, 134–6, 147
Botswana, 17
bottomland hardwoods, I*b*, I*c*, 137
boulder, 140
Brazil, 218
breeding
    area, 155, 168
    behavior, 2, 8, 103
    chronology, 100, 158
    habitat, 93, 102, 158, 174
    range, 88, 91, 101, 175
brood, 32, 41, 140
buffalo, 231
bush, 27, 122, 129
Buttonbush, IV*j*, 123, 136

California, 89, 133, 214, 218, 221
Canada, 5, 16, 53, 57, 88, 89, 107, 113, 133, 140, 156,
    173, 174, 185, 191, 203, 216, 218, 220, 240
carnivore, 8, 93
carp, 9, 218
cattail, 14, 42, 67, 72, 103, 123, 133, 153, 157, 183, 185
cattle, 35, 69, 88, 114, 221
census, 191, 199, 200, 202, 205

Printed in the United Kingdom
by Lightning Source UK Ltd.
2037